Systems on silicon

Edited by
P.B.Denyer

AN INTERNATIONAL SPECIALIST SEMINAR ON

SYSTEMS ON SILICON

Organised by the

Electronics Division of the Institution of Electrical Engineers

Venue

Alveston Manor Hotel, Stratford-upon-Avon, England

5-9 SEPTEMBER 1983

Published by: Peter Peregrinus Ltd., London, UK.

© 1984: Peter Peregrinus Ltd.

All rights reserved. No part of this publication may be reproduced, stored in a retrieval system or transmitted in any form or by any means — electronic, mechanical, photocopying, recording or otherwise — without the prior written permission of the publisher.

While the author and the publishers believe that the information and guidance given in this work is correct, all parties must rely upon their own skill and judgment when making use of it. Neither the author nor the publishers assume any liability to anyone for any loss or damage caused by any error or omission in the work, whether such error or omission is the result of negligence or any other cause. Any and all such liability is disclaimed.

ISBN 0 86341 020 0

Printed in England by Short Run Press Ltd., Exeter

IEE DIGITAL ELECTRONICS AND COMPUTING SERIES 4
SERIES EDITORS: S.L. HURST
M.W. SAGE

Systems on silicon

Previous volumes in this series

Volume 1 Stochastic and deterministic averaging processes
P. Mars and W.J. Poppelbaum
Volume 2 Semi-custom IC design and VLSI
P. Hicks (Editor)
Volume 3 Software engineering for microprocessor systems
P.G. Depledge (Editor)

The Institution of Electrical Engineers is not, as a body, responsible for the views or opinions expressed by individual authors or speakers.

Contents

Foreword, G F VANSTONE

SESSION 1 'SYSTEM DESIGN METHODOLOGY'

Chairmen's report, J B G ROBERTS and H J WHITEHOUSE	1
1.1 The impact of VLSI on digital system design, J B CLARY and D W BURLAGE	2
1.2 The impact of VLSI on the architecture of signal processing systems, J B G ROBERTS	3
1.3 Reduced instruction set multi-microcomputer system, L WANG, L FOTI and P C TRELEAVEN	7
1.4 The changing role of analogue and digital processing, C P ASH and K R THROWER	11
1.5 Application of emerging gallium arsenide digital device technologies to the design and fabrication of signal processors operating in high data rate environments, B K GILBERT, B A NAUSED, S A HARTLEY, W K VAN NURDEN and A FIRSTENBERG	16
1.6 Simulation from system design to chip design, K G NICHOLS	20
1.7 Hierarchical data capture: linking systems and chips, P BLACKLEDGE	24

SESSION 2 'CAD TOOLS'

Chairman's report, G F VANSTONE	28
2.1 Design system architecture and the role of of database, P J RUSSELL, N WINTERBOTTOM, S NEWBERRY AND J M SNYDER	29
2.2 Chip design using automatic placement and routing, T YANAGAWA	33
2.3 Hardware specification: a use for hardware description languages?, J D MORISON, N E PEELING and T L THORP	37

2.4 VLSI testing - its potential and problems, T W WILLIAMS — 43

2.5 Testability analysis, W L KEINER — 44

SESSION 3 'PARALLEL WORKSHOP SESSIONS

3.1 Expert systems and VLSI. Chairman's report, G MUSGRAVE — 47

3.2 High speed VLSI. Chairman's report, F G MARSHALL — 49

SESSION 4 'CHIP DESIGN METHODOLOGY AND TECHNOLOGY

Chairmen's report, P B DENYER AND J B CLARY — 50

4.1 Silicon compilers, J P GRAY — 51

4.2 Review of architecture and their interaction with system performance and design methodology, J FOX and S JAMIESON — 52

4.3 Factors limiting performance in minimum geometry devices and circuits, G J DECLERCK — 57

4.4 The technology choice: CMOS v NMOS vs bipolar, D L GRUNDY — 61

SESSION 5 'SYSTEM AND PRODUCT APPLICATION AREAS

Chairmen's report, H J NEWMAN and W FAWCETT — 65

5.1 An overview of emerging markets and requirements for custom/semi-custom VLSI in the commercial market, M G PENN — 66

5.2 Approaches to telecommunications systems integration in silicon, R A BROOMFIELD and A AITKEN — 70

5.3 Personal, low cost, computer systems, J B TANSLEY — 80

5.4 Integrated circuits in consumer products, J D LEGGETT — 33

5.5 Future requirements for VLSI in the military field, K W GRAY and B J DARBY — 85

5.6 Military systems, W S BARDO — 89

5.7 Government applications of a small VLSI facility, E J SPALL — 91

Open session papers — 92

Foreword

The Alveston Manor Hotel, Stratford-upon-Avon, provided a charming atmosphere for the first IEE International Specialist Seminar on Systems on Silicon. This Seminar followed the successful format of four previous Specialist Seminars on Advanced Signal Processing held in Aviemore in 1968, 1973 and 1976 and in Peebles in 1979.

The Organising Committee was pleased that the mix of engineers, managers and scientists selected by personal invitation was a successful one. From the reception in the bar on the first evening to the close of the Seminar participants seemed to fill the hotel and its gardens with lively debate and discussion. This continued in the formal sessions which stimulated much participation from the audience, as appropriate for such a Seminar.

The Seminar was organised around four main sessions: "System design methodology", "CAD tools", "Chip design methodology and technology" and "System and product application areas". For each session a range of presentations was invited and plenty of time allocated for debate at the end of each session. In addition to these sessions, two evening workshops were held; one on "Expert systems and VLSI", and one on "High speed VLSI". Both of these workshops attracted a high attendance and a stimulating level of debate continued, ignorance being no excuse for not participating.

These formal arrangements were complemented by a relaxing social programme. On Wednesday afternoon participants took some time to explore a little of the beauty and history of the surrounding Cotswolds countryside. Many took advantage of an excursion to Warwick Castle. A high point in the programme came with the performance of Twelfth Night at the Royal Shakespeare Theatre on the bank of the River Avon opposite the Alveston Manor Hotel. The play was enjoyed enormously; especially by the overseas visitors. Throughout the week the wives of some participants made many interesting trips to nearby market towns.

This first IEE International Specialist Seminar on Systems on Silicon was altogether a successful venture. The Seminar Organising Committee would like to express its sincere thanks to Mr Andrew Wilson of the IEE for his hard work and patience, and to the session chairmen for their invaluable support.

Dr G F Vanstone

Seminar Organising Committee

Dr G F Vanstone (Chairman)
Mr D D Buss*
Mr J B Clary*
Dr P B Denyer
Mr C H Garnett

Professor Dr H M Lipp*
Mr H J Newman
Professor J Mavor
Professor G Musgrave
Dr J B G Roberts

*corresponding member

List of Authors

AITKEN, A	MARSHALL, F G
ASH, C P	MORISON, J D
BARDO, W S	MUSGRAVE, G
BARRON, I M	NAUSED, B A
BLACKLEDGE, P	NEWBERRY, S
BROOMFIELD, R A	NICHOLS, K G
BURLAGE, D W	PEELING, N E
CLARY, J B	PENN, M G
DARBY, B J	ROBERTS, J B G
DECLERCK, G J	RUSSELL, P J
DENYER, P B	SNYDER, J M
FAWCETT, W	SPALL, E J
FIRSTENBERG, A	TANSLEY, J B
FOTI, L	THORP, T L
FOX, J	THROWER, K R
GILBERT, B K	TRELEAVEN, P C
GRAY, J P	VAN NURDEN, W K
GRAY, K W	VANSTONE, G F
GRUNDY, D L	WANG, L
HARTLEY, S M	WHITEHOUSE, H J
JAMIESON, S	WILLIAMS, T W
KEINER, W L	WINTERBOTTOM, N
LEGGETT, J D	YANAGAWA, T

CHAIRMEN'S REPORT - SESSION 1 "SYSTEM DESIGN METHODOLOGY"

J.B.G. Roberts* and H.J. Whitehouse**

*Royal Signals and Radar Establishment, Malvern, UK, **Naval Ocean Systems Center, USA

The session began with six papers which considered VLSI needs as defined from the system viewpoint, and showed the ways in which both the fundamental features of VLSI and the development of design methodologies must react back on the way in which these needs are formulated. It became clearer that the term VLSI is used to cover too wide a span of complexities so that during the meeting the impression was alternately created and destroyed that the VLSI age is already with us, or that break-throughs in design methods, simulation and interconnection (even more than in fabricating small feature sizes) are needed before VLSI can be seriously applied. If VLSI has over 5,000 gates per chip, perhaps we might suggest ULSI for 50,000 gates minimum.

Both computing and the more specialised field of digital signal processing were examined, and it was found essential for each to use parallel machine architectures. In the case of signal processing, the elements of the system may be fixed function systolic nodes, slave processors or, for some high level tasks such as in pattern recognition, independant computers. The choice between the first two depends on compromising throughput capacity (per unit of hardware) with programmability. Locally programmable processing elements may eventually have to replace slave elements in order to overcome timing problems in broadcasting instructions to them.

Future VLSI computers were certainly seen as being based on multi-processors and much research has explored the language features necessary to exploit parallelism in a large machine, the architecture presumably being conceived as an intimate combination of hard and software. It seems to be necessary to build and experiment with different machine types used on various algorithms before their merits can be reasonably compared. A 'parallel control flow' approach, using reduced instruction set microcomputers, was chosen for the particular work described but the impression left was that architecture designs in this field are more pragmatic than in signal processing, where the structuring of the problems helps to guide the choice.

The declining number of vital signal processing functions for which analogue techniques are still needed was delineated with reference to several classes of spread spectrum radio receiver. Attainable combinations of wordlength and sample rate in A/D conversion remain the principal bar to all-digital radios, although programmable switched capacitor filters still compete very effectively with digital spectrum analysers in FSK.

The threat to silicon based technologies posed by GaAs was assessed. Lower power rail voltages and smaller logic swings are possible with a smaller band gap so that silicon might be beaten in chip dissipation limited situations. However, it is raw gate speed with relatively small chip complexity, applied to those problems in signal processing which preclude great parallelism which seem to hold out a niche for GaAs. However, although analogue device needs will ensure continuous development of the technology, there do seem to be enough problems to prevent its encroaching on digital silicon territory, except as a last resort for specialised applications.

The two papers of session 1 dealing with computer simulation of systems designed for VLSI revealed a large catalogue of unfulfilled wishes. Formalised, machine processable descriptions of function and structure which can be animated within feasable computer speed and memory limits do not exist except for systems smaller than about 300 nodes. Approximate methods are unsatisfactory. For example using functional macros to describe system blocks forces choices between worst case and typical tolerances in their parameterisation. Simulation presently seems to loom as a major obstacle to achieving the potential of digital systems on silicon.

THE IMPACT OF VLSI ON DIGITAL SYSTEM DESIGN

J.B. Clary* and D.W. Burlage**

*RTI, North Carolina, U.S.A., **US Army Missile Command, U.S.A.

ABSTRACT

In this era of systems on silicon, treditional design approaches have been found lacking. No longer is it sufficient to seperate system operational requirements from the integrated circuit design process. A fully integrated design methodology which begins with formal system requirements and ends with verifiably correct systems solutions is needed.

The state of system design practice today is a long way from being fully integrated. In digital systems, for example, boundaries between software and hardware design exist only because we are trained as electrical engineers or computer scientists. Moreover, most of today's hardware design approaches tend to seperate integrated circuit mask level layout from system functional design. To the detriment of system availability, system test approaches are often an afterthought. Even the notion of software running on "virtual machines", while helpful from a software programmer's viewpoint, promotes a lack of sensitivity and understanding of the underlying hardware structure and, perhaps more important, the fundamental relationships that exist between them.

New design approaches, such as the silicon compiler, offer some hope but, at their best, may only be a partial solution. More comprehensive computer-aided design approaches may be the answer, but this is not likely unless they evolve from a technological environment that is truly interdisciplinary.

The advent of VLSI technology and forums such as this one on "Systems on Silicon" could be a starting point for truly integrated systems design. This talk will provide a background for understanding the problem and propose a potential solution.

THE IMPACT OF VLSI ON THE ARCHITECTURES OF SIGNAL PROCESSING SYSTEMS

J.B.G. Roberts

Royal Signals & Radar Establishment, Malvern, UK

INTRODUCTION

As LSI has developed, not only has the level of ambition possible for real time digital signal processing (DSP) increased, but the changed balance of limitations and strengths of the technology has altered the optimum architecture we should apply to a given problem. For instance the availability of compact high speed memory encourages the use of look-up tables and two-dimensional processing of images. Further, the VLSI constraints react back on the algorithms used so that, for instance, a 32-point DFT may be easier to implement than an in-place FFT algorithm because control functions can be more costly than arithmetic or memory.

Shrinking the feature size of VLSI by a factor K applied blindly will not produce the orders of magnitude improvement which might be expected. Predictions of a K^3 improvement in power-delay product and gate-Hz per cm^2 of silicon ignore the problem of matching the gate delay improvement with the transmission line interconnections on chip, and the chip area consumed by drivers for inter-chip signalling. It is vital to keep the signalling distances short and systematic. The conclusion here is that 'the medium is the message.' We build chairs out of wood, bridges out of steel and digital processors out of planar silicon. In each case we need to work with the 'grain' of the material exploiting its virtues to the full.

It is interesting that the range of signal processing requirements shows such commonality at the low functional level. The use of FFT's, filtering, adaptive thresholding, integration etc is universal and the major differences in processing requirements are not so much between say sonar, radar and image processing but between highly flexible miniaturised manpack equipments, high throughput specialised processors (eg missile homing heads) and large ground or ship-based installations where size and power constraints are relaxed but high throughput must be combined with software-determined flexibility. Although there is no unique way of subdividing the field we can aim to cover most needs using a small set of well chosen processor types with acceptable costs because each has a reasonable range of system applications.

Let me therefore propose three types of signal processor each exploiting VLSI in a somewhat different way and satisfying its own class of requirement. The hope is that almost any requirement can be satisfied by one of them and that as technology continues to improve, the architectures and much of the software will be stable whilst bandwidth, number of channels processed etc. will increase. I shall also mention briefly two other developments which look interesting and may change the scene in due course.

SIGNAL PROCESSING MICROPROCESSORS

Already an important range of needs can be met using a single chip processor specialised towards the sum-of-product operations on data arrays basic to correlation, transforms, filtering, beamforming etc. At the present level of technology this class of chip uses fixed point arithmetic and has a hardware multiplier occupying a significant fraction of the chip area, a conventional ALU and an on-chip data RAM of a few hundred words, distinct from the program store which is ROM either on or off-chip. The RAM is organised to deliver pairs of operands to the multiplier at each clock cycle and to accept new data by DMA without interrupting processing. Data address generators are available to scan array indices in normal and bit-reversed order making filtering and FFT operations simple and efficient to program.

Representative chips of this type are the Texas Instruments TMS 320 and the STL DSP 128.

The main parameters of the latter are listed in Table 1 together with some performance benchmarks.

We can regard these chips as signal processing variants of the conventional micro: the instruction set allows a wide range of algorithms to be implemented but the processor is maximally efficient for algorithms dominated by fixed point multiply-add operations on arrays of data. For such roles as versatile modems, digital filters, correlators, and spectrum analysers at telecommunications bandwidths, this type of essentially single chip processor offers a cost effective utilisation of VLSI: a standard part, software-adaptable to many needs. For high volume applications, the program can be incorporated in an on-chip ROM during manufacture. This current architecture seems likely to survive technology updates whilst its speed, storage capacity and wordlength (probably in floating point) will of course improve. 2-3 MOP/s and ~1K byte of onboard RAM are possible in current technology and will probably improve by an order of magnitude with ~1µm feature size thus reinforcing the utility of a single chip processor.

HIERARCHICALLY CONTROLLED MODULAR PROCESSORS

The appearance of the TRW fast multiplier and multiply-accumulate chips a few years ago exerted a profound influence on DSP. They represented an optimum use of contemporary IC technology offering an almost universally required signal processing function in a single chip. These devices allow little programmability: products can be added or subtracted, the output may be single or double precision and may be rounded. Processors using them

Multiplier	:	16 x 16 bit → 32 bit every 400ns
ALU	:	35 bit
On chip data RAM	:	512 x 16 bits
Program memory	:	external ROM, up to 64 Kbytes
Technology	:	3μ NMOS
200 tap transversal filter	:	82μs update
64 tap complex equaliser	:	106μs
50 tap lattice filter	:	121μs
10 second order section IIR filter	:	30μs
64 point complex FFT	:	1.9ms

Table 1   Performance of the DSP 128 processor

can either be hard wired or incorporate the multiplier chips into somewhat larger functional units, eg 32 point DFTs or filter sections, several of which can be assembled to form a processor controlled and monitored by standard microprocessors. The system hardware modules are then drawn from a small inventory numbering perhaps 5 types. Each is specialised to a particular function, perhaps beamforming or bandshifting and filtering or spectral estimation but has parametric flexibility selectable by the system control software.

Currently such system modules exist as cards often implementing functions in block floating point using LSI chips providing fixed point arithmetic, store, address counters, etc. Again this system structure is likely to benefit from VLSI advances rather than be superseded: the chip functions will increasingly be devolved to macrocells within chips and the cards will shrink as the number of packages required decreases. Pin connections, power consumption and speed will all benefit. Also hardware efficiency, economy and low design risk result from customising a system from pre-designed reliable and replaceable units. Self checking and substitution of spare modules can be incorporated and the system specification and probably the software can survive technology updates.

The modular processor is particularly well matched to application areas where several related systems are envisaged (eg radars or sonars with different numbers of beams or doppler channels) and also where hardware compactness is important, because specialised modules can be developed to meet very intensive tasks such as adaptive beamforming or associative memory searches for fast signal sorting.

Completely dedicated, fixed function hardware with custom VLSI would of course offer the ultimate in size and power consumption with no flexibility to adjust to changing scenarios; and a software-driven general purpose processor is at the other extreme. The modular processor offers the best compromise for many applications. Current card modules offer ~2 x 10^6 block floating point real multiply-add operations/sec whilst chips for the US VHSIC program in 1.25μ NMOS are beginning to offer 2 to 4 times this capability.

However, as system complexity increases and more algorithmic flexibility is demanded, problems can begin to arise in scheduling and handling the data flow. Control overheads appear with segmented bussing, data routing switches or tagging data to implement 'data flow' scheduling and the efficient generation of microcode, possibly automatically from a high level language, appears to be difficult for a complex system which is required to be versatile also. These problems invite us to consider a third class of VLSI signal processors.

DISTRIBUTED ARRAY PROCESSORS

The outstanding feature of digital VLSI is its capacity to replicate elements of logic and memory with very high complexity, density, and gate speed, both within chips and by reproducing identical chips. What is difficult, is to perfect a new design involving many disparate sub-functions interconnected in a pseudo random way, and to utilise the potential gate speed without being limited by transmission line effects and race conditions. There are therefore technological incentives to use processor architectures based on large scale replication of simple structures with regular and short interconnection paths for data.

A more fundamental thrust in the same direction is provided by the realisation that high speed signal processors and supercomputers are 'tying one hand behind their backs' by relying on higher technology versions of Von Neumann machines with very limited pipelining and parallelism. Cutting loose from serially organised algorithms and hardware poses significant problems in computing however because the end user may wish to perform calculations which do not admit of parallel processing, or he may be unwilling to recast into parallel form, software which may result from years of work in say weather forecasting. Signal processing however is overwhelmingly parallel in nature with great scope for pipelining through the largely data-independent algorithms.

Thus architectures of the ilk of ILLIAC IV, CLIP, DAP, MPP and GRID which concurrently use thousands of identical bit-serial processing elements (pe's) with nearest neighbour data interconnections in two dimensions, have come to the fore with the advent of VLSI, for large signal processing problems. Each pe accesses a local memory of several kbits and, under the control of a central Master Control Unit (MCU in Fig 1), executes the same program instructions as every other pe (SIMD: Single Instruction on Multiple Data elements, except that the rewrites to memory can be selectively inhibited when required).

Some of the modular features of our previous category carry over (scope for redundancy and for building large and small processors from the same VLSI elements) but the main objective is quite different: to provide a compact, well specified machine with established program development software, library routines and high and low level languages. A machine like this, the Distributed Array Processor (DAP) described by Hunt and Reddaway (1) has been on the computer market for some years but is unsuited to real time signal processing in lacking fast I/O, requiring a host machine and occupying several cabinets. However, even current gate array and RAM technologies allow a 1000-pe version with 2M bytes of memory and a data-organising I/O buffer within a volume of 1 cu ft and consuming a few hundred watts. The power of such a machine is not readily summarised in MOP/s because this depends on the word lengths in use which may be 1-bit for logic operations or as long as required during fixed or floating point computation. However some benchmark speeds for a machine to be delivered in 1984 are:-

8-bit integer additions	300 MHz
32-bit integer additions	80 MHz
32-bit floating point additions	10 MHz
32-bit floating point multiplications	6 MHz
32-bit floating point square roots	8 MHz
16-bit fixed point 32 point FFT	200 KHz
16-bit fixed point 1024 point FFT	2 KHz

A closely related current development, the GRID (GEC Rectangular Image and Data Processor) connects 8 rather than 4 nearest neighbours and incorporates additional address generation and histogram extraction features but still conforms to the basic scheme of Fig 1. A crucial feature in applying these processors is the necessity to re-think the algorithms in some detail for processing in a highly concurrent way. High and low level languages are available embodying parallel constructs appropriate to the machine and it is essential to exploit these intelligently rather than to attempt to transfer algorithms optimised for serial execution to the parallel environment.

We are currently assessing the applicability of a current technology 32 x 32 pe signal processing DAP with 1 or 2 M bytes of memory and a clock cycle time in the region of 130 ns to real time military requirements in airborne radar, image processing, speech recognition and ESM. Results to date confirm the power of the processor, not only for rigidly structured arithmetic routines such as FFTs and filtering, but also for less predictable algorithms such as resolving ambiguities in range and doppler, adaptively thresholding two dimensional data, and dynamic programming for time-warped speech processing. It is encouraging to find that the full parallelism of the machine can be brought to bear by sensible structuring of the data format in all the cases studied. The days of the present 'array processor' of very modest parallelism and pipelining may be numbered once parallel programming becomes accepted because of the efficiency with which highly parallel machines can utilise VLSI.

An advanced technology version of the distributed array processor is likely to show considerably enhanced performance because of the systematic connectivity as we pack more pe's and store per chip. The architecture is still dominated by technology issues and my instinct is to regard the processor as interconnected memory modules with a relatively small logic capability integrated into each. However, there are several live issues: for instance whether the pe's should remain bit serial or become bit slice and whether the memory should be a fast cache + slower dynamic RAM hierarchy.

TWO OTHER OPTIONS

My tidy picture implying that three classes of signal processor architecture realised in succeeding generations of VLSI could between them cover most signal processing requirements, is always liable to be improved upon by new ideas. At present two such possibilities are being researched: systolic arrays and the transputer.

The systolic array (SA) concepts of Kung and Leiserson (2) have prompted considerable interest in mechanising complex numerical algorithms through iterative steps executed by systematically clocking data, partial results and coefficients through a one or two-dimensional lattice of connected processing elements. The philosophy is very close to that of the lockstep processing and nearest neighbour data transfers of the SIMD machine except that in SA the same processing step, eg multiply and accumulate is not only replicated everywhere within the processor (except possibly at the boundaries) but also remains fixed in time. Control overheads thus vanish to be replaced by clocking waveforms.

Computation-intensive tasks such as matrix multiplication and least squares minimisations (3) could be devolved to a SA peripheral hosted by a more conventional computer. Alternatively an SA module might be incorporated into a modular signal processor. These possibilities remain whilst the fruitful SA concept can also be applied at bit level for single chip multipliers and convolvers (4,5). The numerical algorithms are equally valid for distributed array processors.

Fisher and Kung (6) propose programmable pe's at the SA nodes. This suggestion is reminiscent of our discussion of how complex the pe should be in the distributed array processor and how memory should be integrated with it. It is even closer to the concept of the transputer, the eagerly awaited programmable processor-plus-store chip designed with easy facilities for communication between multiple processes being executed in a multi-processor array. The generality of such a machine goes beyond the SIMD capability I have cited as appropriate for signal processing, since each processor is independently programmed. However, scheduling the processing may be made tractable and efficient by a simplified instruction set and the use of a new programming language, OCCAM, which allows the programmer to explicitly designate instructions which need to be sequential and those which may use parallelism available within the hardware. The role for the transputer in signal processing is of course yet to be evaluated so judgement must be very tentative for the moment.

THE NEED FOR CUSTOM DESIGN VLSI

In proposing a small spectrum of signal

processing architectures, I have not implied that no other options are possible. Quite the reverse, I see it as cost effective to limit the potentially infinite scope for piecemeal developments and to discipline the development of VLSI components and to make available well specified chip sets, macrocell designs and processors together with support software, so that design delays can be minimised and costs spread over larger production runs.

I am conscious of not having placed a high premium on semi-custom design methods, silicon compilers and other methods for achieving fast design and fabrication turnround. Except for the use of standard macrocells in modular processors, I foresee a rather limited need for low volume custom VLSI in signal processing because, where customising is appropriate, for instance in tailoring a specialised processor into the confines of a missile head, or achieving very low power consumption for a miniature remote sensor; production runs are longer and the system design is probably protracted in any case. My plea is therefore for industry-standard parts (and software) manufactured in high volumes and made generally available at low cost.

REFERENCES

1. Hunt, D.J. and Reddaway S.F., 1983, "Distributed Processing Power in Memory", Chapter 5 in "The Fifth Generation Computer Project", G.G. Scarrott (Ed), Pergamon Infotech Ltd, Maidenhead, UK.

2. Kung, H.T. and Leiserson, C.E., 1978, "Systolic Arrays for VLSI Processor Arrays", reprinted as section 8.3 of "An Introduction to VLSI Systems", Mead, C. and Conway, L., Addison-Wesley, 1980.

3. McWhirter, J.G., 1983, "Recursive least-squares minimization using a systolic array", Proc SPIE Real Time Signal Processing VI, San Diego 1983.

4. McCanny, J.V. and McWhirter, J.G., 1982, "Implementation of Signal Processing Functions using 1-bit Systolic Arrays", Elec Lett 18, No 6, pp 241-243.

5. McWhirter, J.G. and McCanny, J.V., 1982, "A Novel multi-bit convolver/correlator chip design based on systolic array principles", Proc SPIE Real Time Signal Processing V, Arlington.

6. Fisher, A.L. and King H.T., 1982, Proc USC Workshop on VLSI and Modern Signal Processing, University of Southern California, Los Angeles, 1-3 November, (In the press).

© Controller, HMSO, London 1983

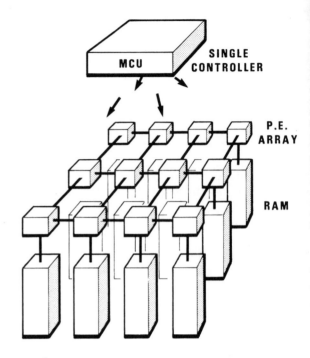

Fig 1: Outline architecture of the ICL DAP

PROGRAM ORGANISATIONS FOR MULTI-MICROCOMPUTER SYSTEMS

Philip Treleaven, Lewis Foti and L. Wang

University of Newcastle upon Tyne, UK

Attempting to make larger-scale single microprocessors in VLSI scaled to submicron dimensions becomes self-defeating due to communications problems. One obvious solution is miniature microcomputers that can be replicated like memory cells and operate as a multi-processor architecture. (Such multi-processors may form the basis of a new VLSI generation of components.) The fundamental problem to be solved is how to orchestrate a single computation so that it can be distributed across such an ensemble of processors. Various parallel program organisations might be used: data flow, reduction, actor, logic or even control flow. This paper assesses these parallel program organisations, concluding that control flow is the most primitive and fundamental program organisation. It then presents the design of the Reduced Instruction set Multi-Microcomputer System (RIMMS) which is based on "parallel" control flow.

DESIGN PHILOSOPHY

Traditionally, the trend in designing microprocessors and mainframe computers has been towards increasingly complex instruction sets and associated architectures [4]. In contrast, designs based on the so-called reduced instruction set [3,6] philosophy have a simple set of instructions, and a correspondingly simple machine organisation tailored to the efficient execution of these instructions. However, attempting to make larger single microprocessors in very large scale integration (VLSI) scaled to submicron dimensions becomes self-defeating, due to communications problems and the escalating costs of designing and testing such complex processors [8]. One obvious solution is miniature (reduced instruction set) microcomputers which can be replicated like memory cells and operate as a multi-processor system. The fundamental problem to be solved is how to orchestrate a single computation so that it can be distributed across such an ensemble of processors [5].

Various parallel program organisations [7] might be used: control flow, data flow, reduction, actor, and logic. These organisations are distinguished by the way data is communicated between instructions, and by the way one instruction causes the execution of others.

In this paper we discuss these parallel program organisations and then present the design of the Reduced Instruction set Multi-Microcomputer Systems (RIMMS) which is based on "parallel" control flow.

The aim of the ongoing RIMMS project is to design a simple conventional microcomputer with primative communications mechanisms that is able to form a component of a tightly-coupled multi-microcomputer system.

A RIMMS microcomputer has a 16-bit word size, with each register, data element and address being 16 bits. Instructions are 2 words (32 bits) and use a 3-address format. There are less than 20 operators. Each microcomputer in the multi-microcomputer system is addressable, and behaves as a combined memory and processor that is able to service load, store and execute operations. The 16-bit global address consists of two parts: the high 8 bits define a specific microcomputer, while the low 8 bits define a word in that microcomputer's memory. Although a microcomputer can access any word in the global address space, an attempt to execute alien code causes execution to transfer to the specified microcomputer.

This design contains a number of important concepts. Firstly, although a microcomputer can make a data access to any word in the global address space, code is always executed by the local microcomputer (to encourage locality). Secondly, a microcomputer has the ability using a FORK instruction to create a parallel flow of control. Thirdly, a microcomputer executes a process to completion thus providing a primitive form of synchronised access to the contents of its local memory. Finally, to enable simple process migration the amount of state information held in the processor's registers is minimised.

PARALLEL AND DISTRIBUTED COMPUTING

There are basically five categories of program organisation (shown in Figure 1) [7] on which a multi-microcomputer system could be based. Each with an associated category of programming languages. Firstly there are control flow organisations and procedural languages. In a control flow organisation explicit flow(s) of control cause the execution of instructions. In a procedural language (e.g. BASIC, FORTRAN) the basic concepts are: a global memory of cells, assignment as the basic action, and implicit (sequential) control structures for the execution of statements.

Secondly there are data flow organisations and single-assignment languages. In a data flow organisation the availability of input operands triggers the execution of the instruction which consumes the inputs. In a single-assignment language (e.g. ID, LUCID, VAL, VALID) the basic concepts are: data "flows" from one statement to another, execution of statements is data driven, and identifiers obey the single-assignment rule.

Prog lang	procdl.	single-assign.	applicat.	object-oriented	pred. logic
comp arch	control flow	data flow	reduction	actor	logic

Figure 1: Classes of Languages/Architectures

Thirdly there are reduction organisations and applicative languages. In a reduction organisation the requirement for a result triggers the execution of the instruction that will generate the value. In an applicative language (e.g. Pure LISP, SASL, FP) the basic concepts are: application of functions to structures, and all structures are expressions in the mathematical sense.

Fourthly there are actor organisations and object-oriented languages. In an actor organisation the arrival of a message for an instruction causes the instruction to execute. In an object-oriented language (e.g. SMALLTALK) the basic concepts are: objects are viewed as active, they may contain state, and objects communicate by sending messages.

Lastly there are logic organisations and predicate logic languages. In a logic organisation an instruction is executed when it matches a target pattern and parallelism or backtracking is used to execute alternatives to the instruction. In a predicate logic language (e.g. PROLOG) the basic concepts are: statements are relations of a restricted form, and execution is a suitably controlled logical deduction from the statements.

As an illustration of these novel parallel and distributed forms of computing we will briefly examine data flow. The most important properties of data flow are that instructions pass their results directly to all the consuming instructions and that an instruction is executed when it has received all its inputs - properties that influence the general-purpose nature of data flow. The operation of a data flow program organisation is shown in Figure 2.

In Figure 2 each data flow instruction consists of an operator, two input operands which are either literals or required data tokens, and a reference such as "i3/1" defining a consumer instruction and argument position for the result data token. Data tokens are used to pass data from one instruction to another and they are also used to cause the execution of instructions. An instruction is enabled for execution when all its input arguments are available, i.e. when all its data tokens have arrived. The operator then consumes the data tokens, performs the required operation, and using the embedded reference stores a copy of the result data token into the consumer instruction(s).

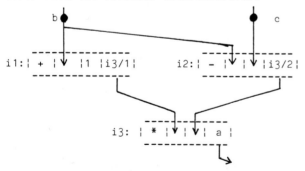

Figure 2: Data Flow Program for a=(b+1)*(b-c)

Data flow systems are most naturally programmed in a very high-level form of programming called single-assignment languages [1]. Single-assignment languages are based on a rule stating: a variable may appear on the lefthand side of only one statement in a program fragment. This allows the data dependencies in a program to be easily detectable and so statements may be specified in any order. As an illustration of single-assignment programming we will examine a procedure in ID [1] for inner-product ai*bi:

```
procedure inner-product (a, b, n)
  ( initial s <- 0
    for i from 1 to n do
        new s <- s + (a[i]*b[i])
    return s )
```

Figure 3: Single-Assignment Program Σ ai*bi

This procedure takes as input two arrays "a" and "b", both of length "n", and returns their inner-product "s". These statements have the following interpretation:

```
       s0 <- 0

       s1 <- s0 + (a[1] * b[i])
       s2 <- s1 + (a[2] * b[2])
       ...
       sn <- sn-1 + (a[n] * b[n])

       return sn
```

and hence obey the single-assignment rule. Since execution is driven by the availability of data, all the multiplications can execute in parallel, after which the tree of partial results will be summed to produce the result "sn".

A problem with the novel parallel program organisations, such as data flow, is that they are largely unproven and, in addition, represent approaches that discard the massive investment in traditional control flow computing. However, our investigations into program organisations [7] reveal that control flow is the most primitive organisation, in that it can be used to simply implement the other four organisations. Thus in the next section we present a multi-microcomputer system based on a parallel control flow program organisation.

RIMMS MULTI MICROCOMPUTER

The architecture of RIMMS is described in terms of two levels of machine: the multi-microcomputer level handles inter-process(or) communication supporting non-local load, store and execute operations; and the microcomputer level services these operations and handles the atomic execution of a single process.

Multi-Microcomputer System

RIMMS consists of a linear array of up to 255 microcomputers that communicate via a shared bus, as shown in Figure 4. Each microcomputer has a simple processor and 256 words of local memory.

```
---------------------------------------------
|       8-bit global address                |
|      ---------------------------          |
|   1:|        2:|      ... 255:|           |
|   ----------   --------   -----------     |
|  | processor| | processo| | processor|    |
|  |----------| |---------| |----------|    |
|  | memory   | | memory  | | memory   |    |
|  |(8-bit local||(8-bit lo||(8-bit local|  |
|  | address) | | address | | address  |    |
|   ----------   --------   -----------     |
---------------------------------------------
```

Figure 4: Multi-Microcomputer System

The system has a 16-bit address space:

```
               address
      global (8 bits) local (8 bits)

      ---------------------------------
      | microcomputer | memory cell   |
      ---------------------------------
```

Figure 5: RIMMS Address

The top 8 bits is a global address (in the range 1-255) defining a microcomputer, while the bottom 8 bits is a local address (in the range 0-255) defining a word in it's memory. (Global address 0 is the default for specifying the current local address space and is therefore not recognised at the Multi-Microcomputer level.)

When one microcomputer wishes to communicate with another, for example to access it's memory, the microcomputer generates a "packet". The format of a packet, as shown in Figure 6, consists of a 2-bit operation field, a 2x8-bit destination address, and a 16-bit operand.

```
                  global  local
     2 bits      8 bits  8 bits     16 bits

   -----------------------------------------
   | operation |    address   |  operand   |
   -----------------------------------------
```

Figure 6: Multi-microcomputer packet format

The packet operations are defined as follows:

LOAD - copies the contents of MEMORY[address] to the microcomputer's register defined by the 16-bit operand. This is implemented by the destination microcomputer generating a STORE_REG packet.

STORE_REG- places the operand in the microcomputer's register defined by the address.

STORE_MEM- places the operand into the MEMORY[address].

EXECUTE- starts a new process whose code is at MEMORY[address] and data environment is at MEMORY[operand].

For all these packets the global address defines the destination microcomputer.

When a packet is sent the destination microcomputer may accept or reject the packet. If rejected, the source microcomputer will attempt to re-send the packet at the next opportunity. Whether a packet is accepted or rejected depends on the status of the processor and memory of the destination microcomputer. In simple terms, load and store operations may be serviced by the memory concurrently with the operation of the processor. However an execute packet may only be accepted when the processor is idle, having completed the execution of its previous process.

Next we examine the architecture of a microcomputer.

Microcomputer

The microcomputer level machine consists of three basic components: the local memory of 256x16-bit words, the memory controller, and the 16-bit processor for arithmetic, as illustrated by Figure 7.

```
                                      bus
                                   ---------
                                       |
   ------------------------------------------
   | Processor        | Memory             |
   | (ALU + registers)| Controller         |
   |----------------------------------------|
   |       l o c a l   m e m o r y         |
   |         256 x 16-bit words            |
   ------------------------------------------
```

Figure 7: Microcomputer

The memory controller is connected to the global bus, and to the local processor and memory. It supports communication, in the form of packets, between these three units. To hold a packet, the memory controller has 3 registers: a 2-bit memory operation register, a 16-bit memory address register, and a 16-bit memory data register (see Figure 8).

```
memory operation register   MOP   ( 2 bits)
memory address register     MAR   (16 bits)
memory data register        MDR     '    '
```

Figure 8: Memory Controller Registers

These registers correspond to the operation, address and operand fields, respectively, of a packet.

When a memory controller is idle it can receive a packet either from the local processor or from some other microcomputer. A packet from the processor can be destined for the local memory or for another microcomputer, whereas a packet from the bus can be destined for the local processor or memory.

The processor, the last component of the microcomputer, consists of an arithmetic logical unit (ALU) and seven 16-bit registers. Each register, data element and address is 16 bits. Instructions, however, are 32 bits and use a 3-address format. Figure 9 shows the 7 registers of which only two are addressable.

```
program counter         C     (16 bits)
data register           D     (16 bits)
instruction registers   I1,I2 (2x16 bits)
ALU register 1          A1    (16 bits)
ALU register 2          A2      '    '
ALU register 3          A3      '    '
```

Figure 9: Processor Registers

C the program counter points to the local code currently being executed. D the data register points to the current data environment which may be any where in the address space. I1,I2 holds the current 32-bit instruction. A1,A2,A3 are the input registers to the ALU, holding the current instruction's operands. Their contents have no meaning from one instruction to the next.

An instruction's format, as illustrated by Figure 10, consists of: a 5-bit operator field, 3x1-bit mode (Mi) fields, and 3x8-bit operand (Oi) fields. Modes and arguments are interpreted as follows. If the value of mode bit Mi=0 then the corresponding 8-bit operand Oi is treated as a literal. Oi is sign extended to 16 bits and the resulting argument is placed in the corresponding ALU register Ai. If the mode bit Mi=1 then the 8-bit operand Oi is treated as a signed displacement relative to the data register D. The resulting address D+Oi is de-referenced (via the multi-microcomputer if necessary) and the memory contents is placed in the ALU register Ai. Notice that the modes and operands are interpreted independently both of the operator and of whether they are to be used for input and output by the ALU. However, the operator does determine how many of the three argument are used by the ALU.

```
         M1 M2 M3   O1      O2      O3
5 bits  1   1  1   8 bits  8 bits  8 bits
-------------------------------------------
|operator|mode bits|lit/add|lit/add|lit/add|
-------------------------------------------
              0 -  literal
              1 -  address (memory [D+literal])
```

Figure 10: Microcomputer Instruction Format

The ALU supports only two information types: 16-bit integers (2's complement) and booleans (TRUE=FFFF, FALSE<>FFFF), and following the reduced instruction set philosophy only a minimal set of operators are provided. These operators are listed in Figure 11.

Operation	Mnemonic	Description
arithmetic	ADD	-
	SUB	-
logical	AND	-
	OR	-
	NOT	-
shift	LSHIFT	logical shift
	ASHIFT	arithmetic shift
compare	EQ	equals
	GT	greater than
control	IF	if TRUE jump
	FORK	fork flow of control
	HALT	halt processor
movement	MOVE	move arg to address
	STORE C	store program counter
	LOAD D	load data register
	STORE D	store data register

Figure 11: Processor Instruction Set

Finally, note that the reason we have chosen a 3-address instruction format and only two addressable registers is to minimise the state information that needs to be moved from one microcomputer to another, when control is transferred.

Implementation

In the initial implementation of RIMMS our aim has been to keep the structures as conventional and conceptually simple as possible. The multi-microcomputer level implementation consists of an 8-bit, passive, bus to which is connected the microcomputers. Access to the bus by the microcomputers is controlled by an additional wire loop, "daisy-chained" through the micros. This wire conveys a single "token", successively from one microcomputer to the next. When the token is received, the microcomputer may attempt to transmit a packet. When a packet is accepted, the packet is transmitted 8-bits at a time between the memory controller registers of the source and destination microcomputers. When the microcomputer finishes with the bus, the token is passed to the next microcomputer. A microcomputer is implemented as three components: a custom data path chip, a custom PLA chip for the memory controller, and commercial chips for the memory.

Any reader requiring further details on parallel program organisations should refer to [7] or on the RIMMS multi-microcomputer should consult [2].

REFERENCES

1. Arvind and Gostelow K.P., "The U-Interpreter", COMPUTER, vol. 15, no. 2 (February 1982) pp. 42-49.

2. Foti L. et al, "Design of the Reduced Instruction set Multi-Microcomputer System (RIMMS)", Computing Laboratory, University of Newcastle upon Tyne, Internal Report (August 1983).

3. Patterson D. and Sequin C., "A VLSI RISC", COMPUTER, vol. 15, no. 9, (September 1982) pp. 8-21.

4. Patterson D. and Ditzel D., "The Case for the Reduced Instruction Set Computer", Computer Architecture News, vol. 8, no. 6, pp. 25-32, October 1980.

5. Seitz C., "Ensemble Architectures for VLSI - A Survey and Taxonomy", Proc. 1982 Conf. on Advanced Research in VLSI, P. Penfield ed., MIT, pp. 33-45, January 1982.

6. Taylor R. and Wilson P.: "OCCAM Process-oriented language meets demands of distributed processing", Electronics (November 1982) pp. 89-95.

7. Treleaven P.C. et al, "Data Driven and Demand Driven Computer Architecture", ACM Computing Surveys, vol. 14, no. 1, pp. 93-143, March 1982.

8. Treleaven P.C., "VLSI Processor Architectures", COMPUTER, vol. 15, no. 6, pp. 33-45, June 1982.

THE CHANGING ROLE OF ANALOGUE AND DIGITAL PROCESSING

C.P. Ash and K.R. Thrower

Racal Research Ltd, UK

INTRODUCTION

Until very recently signal processing has been implemented using analogue, "continuous time" techniques. The active devices have been vacuum tubes or transistors, in conjunction with passive L, C and R elements. The modern trend is towards "sampled-data" techniques where both digital (numeric) processing and "sampled-data analogue" (charge transfer devices, including switched capacitor filters) play significant roles.

Digital techniques are "attacking" on two main fronts. The first is the highly complex system where cost is of secondary importance to performance and flexibility. The second is where low cost and low power consumption are of prime importance, but where performance must be maintained. Typical examples of both are modems, beam forming, spectral analysis, correlators, programmable filters and general purpose processors.

The traditional analogue high frequency filtering techniques are also being challenged by monolithic SAW and Bulk devices. Examples are chirp filters for radar and spectral analysis, correlators for spread spectrum reception and wide band filters for TV.

The future, however, lies with the improvement of digital techniques. These will make significant inroads into almost all areas of analogue signal processing, particularly when frequencies and bandwidths are below 100MHz. On the horizon Ga-As processes, or possibly sub-micron MOS could extend the operating ranges above 100MHz.

THE TREND TOWARDS DIGITAL PROCESSING

Possibly the simplest way of defining how the role of analogue and digital processing is changing, is by inspecting the interface between analogue and digital in a system. This interface is the all important analogue to digital and digital to analogue converters (A/D, D/A), and the trend of the change is twofold. Firstly, there is a continual demand for higher conversion rates to allow processing of wider bandwidths; simultaneously there is a demand for an increase in the conversion accuracy, allowing wider dynamic ranges to be encompassed.

Secondly there is a demand for reduction in size, power consumption and price without any loss of performance.

By looking at the improvements in A/D converters one can see why the modern trend is moving towards digital processing. In general A/D converters can be divided into three main classes (see Table 1).

14-16 bit precision, medium speed

These devices are used where wide instantaneous dynamic ranges (up to 90dB) are required at sampling rates of the order of 100kHz. Applications are restricted to the processing of audio sugnals (in sound studios for example) and critical filtering or correlation applications, where bandwidth is of secondary importance to dynamic range.

10-14 bits, medium to high speed

These devices are used for general purpose application with sampling rates up to 20MHz and where moderate dynamic ranges (60-80dB) are sufficient. Several commercial equipments, using digital processing techniques, based on these devices are now available and many more will be released into production over the next few years. Typical examples of manufactured products at present include line modems and FFT spectrum analysers. In the future, vocoders, radio modems and communication receivers will also be available.

4-10 bit, high speed

These devices are used in low volume and largely military applications where very wide input bandwidths are required, even if this is at the expense of dynamic range. Typical sampling speeds are as high as 100MHz. Applications are in video and radar processing and very wide-band spectrum analysers.

With these modern A/D converters one sees the familiar "push-pull" process, where the push of the new, more advanced, devices are matched by the pull of the potential applications. In recent years, for example, apart from improvements in accuracy and speed, the trend has been towards monolithic devices; this has had a dramatic impact on the price as well as reducing the size and power consumption.

DIGITAL SIGNAL PROCESSING TECHNIQUES

The techniques for digital signal processing have changed dramatically over the last ten years (see Table 2). In the 1970s, it was inconceivable that a general purpose digital signal processor could be built in a realistic way. Digital signal processing was in effect restricted in its applications to "well defined" high volume areas such as line modems. As microprocessors became more widely used for control applications, the spin off into digital signal processing was the microprogrammable bit slice device and, simultaneously, the availability of LSI multiplier chips. The subsequent evolution of these devices over the past five years has been extremely rapid and does not show any signs of slowing. The most recent major step forward is the appearance of single chip digital signal processors with on chip memory, multipliers, arithmetic units and programme sequencers. These devices have started to appear as the processing houses have improved their capabilities of using macro-cell libraries and CAD to give rapid implementation of the system architects' requirements.

Consequently there is no reason why VLSI signal processors cannot become true monoliths with on chip A/D and D/A conversion and memory. The main reluctance to include these devices is the difficulty of specifying the A/D and D/A performance for a wide variety of applications. However, as techniques improve, there is no reason why they should not be over-specified, provided the cost of the devices is not compromised unduly.

ANALOGUE SIGNAL PROCESSING

Before considering further aspects of digital signal processing it is useful to review modern trends in analogue processing techniques - in particular, active filters, switched capacitor filters, charge coupled devices and surface acoustic wave devices. However, none of these will be covered in any significant detail.

Active Filters

The first real break with conventional LC filtering came with the use of C and R devices, combined with integrated circuit amplifiers to realise simple filtering, particularly in the audio frequency range. This technique has now largely replaced the older methods.

Switched Capacitor Filters

A fairly recent innovation has been the truly monolithic replacement for the conventional active filter - the switched capacitor filter (SCF). These are examples of "sampled-data", analogue devices and will be considered later.

Charge Coupled Devices

Charge coupled devices pre-date SCFs and, in some respects, are more versatile. They, too, are "sampled-data", analogue. Almost all devices manufactured have been custom designed with applications including transversal filters, correlators, chirp filters, matched filters, spectrum analysers and imaging. Except for specialised applications, however, their role is now being replaced with true digital processing which can now match the speed, has superior dynamic range and, more importantly perhaps, allow programmability.

Surface Acoustic Wave Devices (SAW)

SAW devices have found their main applications in signal processing where frequencies are in excess of 20MHz - particularly in the 50-1000MHz region. IN these frequency bands, they have been used for IF filters in VHF radio, video filters in TV receivers, narrow-band filters, very wide band filters, correlators, pulse compression for radar, spectrum analysers and even oscillators (when combined with a suitable amplifying device). However, except for very simple applications, the cost is very high. Also the devices have a limited dynamic range (approx. 45dB) and are not suitable for the lower frequencies.

Applications previously only achievable by SAW techniques are now being seriously challenged by digital techniques; this trend will continue as digital devices become faster. Eventually, when gallium arsenide devices become available, digital techniques might easily stretch to 1GHz and, perhaps, beyond.

EXAMPLE APPLICATIONS OF SAMPLED-DATA SIGNAL PROCESSING

As an indication of where sampled-data signal processing is taking over from traditional analogue methods, three examples will be given. The example systems are categorised by their power consumption, which effectively determines their physical size and cost.

Example 1 - Low Power Consumption Filters (< 1 watt)

This is an example of "sampled-data" analogue and is based on switched capacitor filters (Figure 1). The filters are formed in a manner similar to active filters, by combining monolithic amplifiers with capacitors and "resistors". The resistors, however, are synthesised by the use of transmission gate switches and a capacitor. The effective resistance produced by this synthesis is $1/fC$, where f is the clock frequency and C is the capacitance (for example if C = 2pF and f = 100kHz, R = 5M ohms). Quite complex filter combinations can be formed on a single chip. For example one experimental device manufactured by Racal Research Limited has 30 amplifiers, 96 switches and 12 banks of 54 capacitors on a chip of dimensions 0.19 x 0.13 in.

A feature with the SCFs is that they may be programmed in three separate ways - by metal commitment at the last stage of manufacture, by varying the clock frequency and by external programming (for example, by switching internal binary weighted capacitors).

The key features of the SCF are repeatability, stability with temperature, versatility, low power, low cost and compactness. The principal limitations are dynamic range (approx. 60dB) and their relatively low frequencies of operation (<100kHz at present).

Example 2 - Low-Medium Power Consumption Small Rack Mounted Spectrum Analyser

Here we shall consider an equipment with power consumption of the order of 10 watts. This equipment is a low cost spectrum analyser for use by the operator of an HF radio receiver and does not exist in an equivalent analogue form. Typically the parameters required would be:

Displayed Bandwidth - 8kHz, reducible in steps to 500Hz
Horizontal Resolution - Bandwidth/80
Vertical Resolution - 60dB, 40dB, 20dB, with variable baseline

Local Oscillator with frequency readout
Real time Spectral Refresh using FFT analysis

These combinations of parameters would allow an HF radio operator to tune across the band whilst simultaneously viewing, and listening to, the received signals. The advantages of using a digital processor to perform an FFT spectrum analysis are numerous. The most important advantage is the speed of spectral acquisition and true representation of spectra, which is not achieved using an analogue swept filter approach. This means that an operator can easily recognise the difference between transmissions of double sideband, upper or lower sideband voice and data signals such as morse and FSK. These facilities are not possible using a swept filter analyser with a narrow resolution bandwidth. The alternative analogue approach would be to use a filter bank with parallel detectors, but this would make variable resolution extremely difficult to achieve.

Other advantages to be gained by using digital techniques are that the information is already in a form that can be readily converted for use with a liquid crystal display, or the information could be passed to a computer for further analysis.

Example 3 - Medium-High Power Consumption Cabinet Mounting Experimental Equipment

Here we shall consider an equipment which has a power consumption of the order of 500 watts. The purpose of the equipment is to carry out experimental measurements on satellite links by using six distinct receiver types for comparative measurements. The receivers operate at an IF of 70MHz and hence can be used live with suitable up and down converters or can be used in conjunction with a simulator.

The parameters to be measured are as follows:

a) Amplitude and frequency analysis of CW test signals.
b) Carrier Recovery of Phase Modulated signals
c) Analysis of Phase Error spectrum for CW signals
d) Demodulation of BPSK with residual carrier
e) Analysis of search and acquisition for TDMA signals
f) Analysis of carrier recovery and demodulation for BPSK signals
g) Analysis of fast burst acquisition for TDMA signals
h) Analysis of channel state information
i) Analysis of time spreading due to multipath
j) Study of Spread Spectrum acquisition
k) Demodulation and analysis of CDMA spread spectrum signals
l) Analysis of Spread Spectrum synchronisation

To fully analyse these functions the six receiver types are as follows:

1) A PLL Receiver, which will be used to measure parameters a) to e)
2) A BPSK Costas Loop Receiver, which will be used to measure parameters f) to h)
3) A Tuned Filter Carrier Recovery Receiver, which will be used to measure parameters f) to h)
4) A Differentially Coherent BPSK Receiver, which will be used to measure parameters f) to h)
5) A Coherent Spread Spectrum Receiver, which will be used to measure parameters i) to l)
6) A Differentially Coherent Spread Spectrum Receiver, which will be used to measure parameters i) to l).

Figure 2 shows the block diagram of the down-conversion from 70MHz after filtering to two separate IFs of 40MHz and 1.6MHz, followed by analogue crystal filters. This is the analogue circuitry which pre-conditions the signal before Analogue to Digital conversion. The wideband quadrature signals for the spread spectrum receivers, numbers 5 and 6, are converted to 4 bit accuracy at a rate defined by the correlators. The narrowband signal at 1.6MHz is sampled directly at a rate dependent on the receiver number. This direct band-pass sampling technique effectively mixes the signal to a low IF of the order of 10kHz where the digital signal processor can operate on it directly.

The Narrowband Receiver

The first three receivers will not be described in detail as many of their key features are incorporated in the fourth receiver (see figure 3). This is a differential, coherent BPSK receiver, where all the functions are carried out entirely digitally, including the 15kHz oscillator which uses a sine look-up table and the bank of eight narrow band filters at the output.

It is obvious from this diagram that an analogue implementation would be extremely difficult to design, so that temperature effects were negligible, and set-up procedures were simple. The digital implementation has no such problems and in practice performs to within 0.25dB of theoretical in white noise conditions.

The Wideband Receivers

The fifth receiver is a spread spectrum coherent receiver as shown in Figure 4. The spread spectrum correlators are 4 x 1024 bits and can operate on a full length sequence up to 1023 bits or in tiered acquisition mode. The correlator output is provided in both soft and hard decision at twice the chip rate by interfacing into the digital signal processor, which performs the costas loop operations at this rate.

The sixth receiver is a differential coherent spread spectrum receiver as shown in Figure 5. This is an extension of the fifth receiver with the correlators operating in an identical manner with only the digital signal processor changing to operate as in receiver 4.

The use of a programmable digital signal processor has allowed the implementation of all six receivers, and all the options within those receivers, simply by selection of software and data from read only memory. A microprocessor interfaces between the signal processors and the user to provide a front panel on a VDU screen. This allows the user to see the entire state of all selected options at a glance and also allows rapid re-selection of options using simple cursor control. The selection of digital filters is by an address pointer and all filter coefficients are held in the signal processor memory to allow a very rapid access. In general it can be said that very few problems have been encountered in the digital implementation, whereas a purely analogue implementation would have been fraught with drift and set-up difficulties.

CONCLUSIONS

The system designer can see in the near future the VLSI programmable digital signal processing device with on-chip A/D and D/A conversion, programme and data memory. For certain specific military and commercial applications, to-day, low power consumption, single chip, flexible designs are essential. In the very near future miniature commercial communications terminals, typically for mobile telephone communications, and intelligent point of sale transaction terminals will require the adaptability and flexibility which can only realistically be given by digital implementation.

Possibly, more importantly, the increased performance and reduction in high level language will mean less efficient use of the individual VLSI chips, but is essential if the human mind is going to be able to cope with the possibilities of system flexibility and adaptability which must be achievable in the near future.

TABLE 1. Representative range of 4-16 bit A/D Converters

PRECISION	TYPE	SAMPLING RATE	TECHNOLOGY	COMMENTS
16 BITS	Sony CX 899	50kHz	Hybrid	Dual slope integrator
14 BITS	Micro-Networks MN 5260 Zeltex ZAD 2714C	4kHz 100kHz	CMOS Hybrid Hybrid	Low Power Suc. approx.
12 BITS	ADC 85 Micro-Networks MN 5245	100kHz 1MHz	Hybrid Hybrid	Industry standard Suc. approx.
10 BITS	TRW TDC 1013J TRW TDC 1016J	1MHz 20MHz	Bipolar Monolithic	Suc. approx. 0.4Watt Flash 1.5Watt
8 BITS	Sony CX 20052	30MHz	Monolithic	Flash
7 BITS	Philips ADC7C	20MHz	0.7μ NMOS	Flash
6 BITS	AMD 6606	100MHz	Monolithic	Flash
4 BITS	AMD 6688	100MHz	Monolithic	Flash

TABLE 2. Evolutionary development of digital signal processing devices (1970-1982)

DEVICE	DESCRIPTION
PYE TMC 539	Recursive digital filter chip
AMD	4*2 bit multiplier building block
AMI S2811	Mask programmable general purpose digital signal processor
TRW MPY16HJ	16*16 bit multiplier accumulator
AMD 2910 series	Microprogrammable bit slice chip set
FAD	Recursive digital filter chip + detectors
NEC μPD7720	Mask programmable general purpose digital signal processor
TEXAS TMS 320	First 16*16 bit general purpose D.S.P. chip with the capability for external programme and data memory operating at 5MHz clock rate

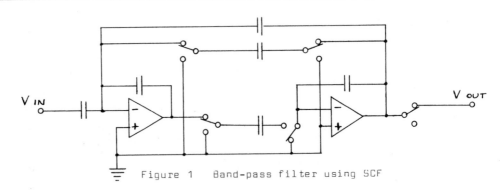

Figure 1   Band-pass filter using SCF

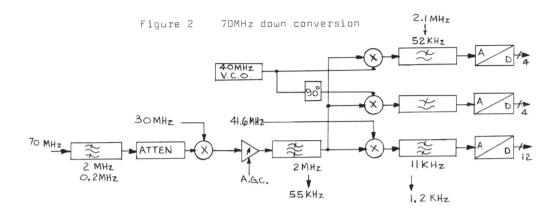

Figure 2   70MHz down conversion

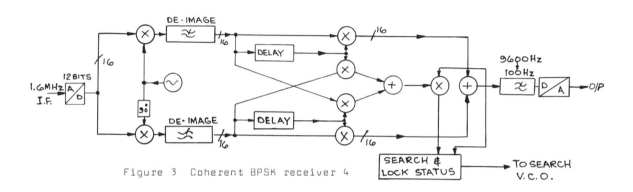

Figure 3   Coherent BPSK receiver 4

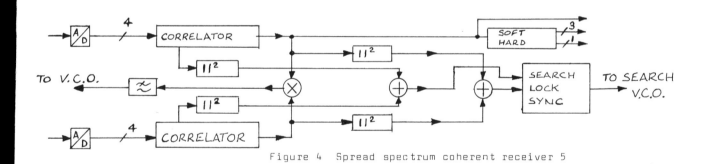

Figure 4   Spread spectrum coherent receiver 5

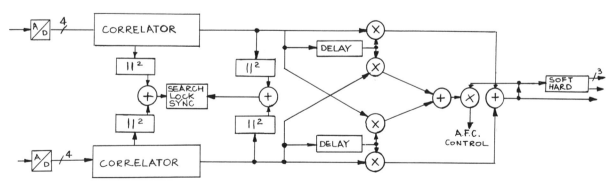

Figure 5   Spread spectrum, differential coherent receiver 6

APPLICATION OF EMERGING GALLIUM ARSENIDE DIGITAL DEVICE TECHNOLOGIES TO THE DESIGN AND
FABRICATION OF SIGNAL PROCESSORS OPERATING IN HIGH DATA RATE ENVIRONMENTS

B. K. Gilbert, B. A. Naused, S. M. Hartley, and W. K. Van Nurden
Mayo Foundation, Rochester, Minnesota

A. Firstenberg
Microelectronics Research and Development Center
Rockwell International
Thousand Oaks, California

ABSTRACT

The emerging application of Gallium Arsenide digital integrated circuits to signal processing problems will require the development of architectures tuned to its special characteristics. Chip design methods may be similar to those used for silicon very high-speed integrated circuit (VHSIC) components, but system design constraints will be unique to GaAs.

INTRODUCTION

Advances in the development of digital Gallium Arsenide (GaAs) integrated circuits have progressed to the point that designers of signal and data processors and of analog conversion modules can begin to discern the systems applications for which they are best suited. It has been commented that Gallium Arsenide may be forced to compete, to its disadvantage, directly with commercial silicon VLSI (greater than 10,000 gates/chip) or military VLSI (referred to as "VHSIC") in a majority of instances. Purely architectural solutions to large signal and data processing problems can often be implemented in VLSI-complexity Silicon integrated circuits, provided that the economic constraints of design costs, unit costs, and schedule can be resolved. More detailed examination indicates that the system implementation tasks fulfilled by Gallium Arsenide integrated circuits will be quite different from and complementary to that of Silicon VLSI and VHSIC. The higher electron mobilities of GaAs transistors result in faster electron transit times across their active regions and hence shorter gate propagation delays for comparable dimensions and power dissipation in silicon devices.

However, because of the lag of the GaAs fabrication process technology behind that of Silicon, the highest gate counts for GaAs components which can be manufactured at reasonable yield levels are currently in the 200-500 gate range, with 1,000-4,000 gate density for logic components (1) and slightly higher gate densities for memory components presently in development and demonstration. Proponents of silicon VLSI technology question the cost effectiveness of assembling systems from components of such "low" device densities. Entire supercomputers are being fabricated from silicon ECL devices exhibiting only SSI densities, and several such machines are being assembled from silicon ECL devices of MSI density (e.g., 2). It could be argued, however, that little additional performance beyond that of the silicon ECL supercomputers would be gained with on-chip gate delays of 100 psec at even MSI device densities, since the chip-to-chip interconnect delays may remain constant and hence dominate system performance. In such a case, a decrease in gate propagation delay at the MSI density level by a factor of 5-10 with respect to Silicon ECL might only improve system performance by 10% or 20%, and at much greater overall system cost as a result of the higher expense of exploiting Gallium Arsenide components. GaAs integrated circuits with densities above the MSI level may thus be necessary; conversely, however, studies performed in the Mayo laboratories indicate that VLSI densities will not be required of the Gallium Arsenide components in order to achieve significant performance improvements (3).

Signal Processing Applications Requiring Digital GaAs

The types of functions which are best suited to Gallium Arsenide must be considered. The majority of near-term applications include high-speed acquisition and perhaps storage of very wide bandwidth pulsed, pseudo-random, or continuous stream data and its processing in real time. Generic areas for the application of these types of processors will be in radar signal processing and signature analysis, electronic countermeasures and electronic support measures, spread spectrum communications, and in selected biomedical environments in which short duration transient events must be recorded (e.g., single molecule reaction kinetics) or very high computational bandwidths are necessary (advanced embodiments of computed tomography; see 4).

The previous comments have stressed the speed performance characteristics of Gallium Arsenide. However, the natural radiation tolerance of GaAs components, in combination with their suitability for very low power operation, makes them ideal for processors which must operate unattended, with a minimum power budget, in hostile environments. Such systems may employ clock rates of only 50-100 MHz; gate power dissipation may be only a few hundred microwatts, and gate delays may be allowed to increase to nearly one nsec.

Design Issues For Digital GaAs

The gate speeds and off-chip signal risetimes of first generation Gallium Arsenide devices suitable for volume production, i.e., in the 100-350 psec regime, are greater than the fastest commercial silicon ECL devices by a factor of 2-5 (Figure 1). Optimum exploitation of these gate speeds will require a different approach to on-chip architectures than for silicon VLSI or VHSIC. As noted earlier, the number of gate delays which occur on each integrated circuit should be maximized, thereby minimizing the system performance loss caused by the interchip connections. A reconsideration of architectural issues will become even more crucial when an advanced generation of GaAs transistors becomes available in several years which can be used to fabricate gates with propagation delays as low as 10-50 psec. The best mechanisms for exploiting such outstanding transistor performance for logic, arithmetic, and memory must be ascertained.

Logic and Arithmetic

The on-chip architectures employed for logic and memory components must be completely reconsidered to maximize their performance while minimizing the amount of system performance dissipated in the intercomponent propagation delays. Simple flow-through logic, e.g., 8:1 multiplexers with two or three levels of gates, 4-bit carry lookahead adders with four gate delay stages, etc., will not make efficient use of the GaAs transistor capability, particularly with the advent of the advanced generation GaAs transistors. Rather than exploiting highly parallel on-chip architectures as is encouraged for VLSI, designers should maximize the number of gate stages traversed by the data operands as they propagate across the component. One effective method of increasing the number of gate stages on-chip is through iterative implementations of signal processing algorithms commonly fabricated as parallel designs (see, e.g., 5). A serial implementation can often be identified by rewriting the signal processing equations in a bit-expanded form, and then operating on the least significant bits of the data operands, followed iteratively by operations on the next most significant bits, etc. Serial or iterative implementations of signal processor algorithms have been only cursorily investigated during the past decade with the advent of ever higher levels of integration in silicon, following a very active period of research during the 1950s and 1960s when high density circuitry was unavailable. Iterative implementation of floating point arithmetic has also been demonstrated, and would be a promising vehicle for the attainment of extremely high throughput in Gallium Arsenide.

Pipelining Techniques

One of the techniques of longest standing for the acceleration of throughput in a signal processor is the addition of data holding registers at intervals throughout the length of its data flow path, i.e., the "pipelining" of the processor. By this means new data operands can be input to the processor before the previous set of operands has completely traversed the processing path. Pipelining techniques are very well known and have been employed since the early days of computer design.

However, in the design of high clock rate processors, there is another use for pipeline registers which has not been exploited previously. It has been frequently stated that interconnect delays between integrated circuits will become a significant fraction of the total clock period as the clock rate rises, thus compromising system performance. The assumption implicit in this observation is that a signal must propagate from the output of a pipeline register, through one or more components and the connections between them, to the input of the next pipeline register, within one clock cycle. In such a design, interconnect delays will indeed comprise a major portion of each clock cycle and degrade system performance. Conversely, it is argued, components must be unacceptably close to one another (maximum separations of 1-2 inches) to yield reasonable performance.

The requirement for inordinately tight packing could be alleviated by designing high clock rate processors such that <u>every</u> integrated circuit contains pipeline registers at both its inputs and outputs. The interconnects created in this manner, with registers at both their source and destination ends, may be conceptualized as pipeline processing stages with a "signal transfer" function; i.e., the wires serve to transfer the signal between two components. The effect of this subterfuge is to allow the wire delay losses to be "pipelined out" of the processor's performance, and also to allow interconnect delays to be virtually as long as the clock cycle. For example, using the wire delay pipelining concept, assume that a signal wavefront propagates at 1 nsec/foot in a printed wiring board of a given dielectric constant. For a system clock rate of 1 GHz, all interconnects could be as long as one foot without degrading system performance at all; at 2 GHz, interconnects could be six inches long, and so on. The concept of wire delay pipelining will become more attractive as GaAs-based signal processors with high clock rates enter design.

Global Architectural Approaches

At a more global architectural level, certain types of multiprocessors appear promising for implementation in GaAs because of their modularity, the amount of processing executed in each node of the multiprocessor, the iterative nature of the processing executed in each node, or the flow-through behavior of the system. These architectures, which include data flow machines, certain types of pipelined FFT processors, array-element-per-pixel image processors (6), and systolic arrays (7), are currently under study in this laboratory.

Arithmetic chip architectures combining iterative implementations with the fast gate speeds of GaAs will probably employ simple microcontrol engines with small on-chip microcode ROMs to control the iterative execution. Microprogram sequencers similar to the AMD2910 will be required, as will the ability to generate and verify the microsubroutines which will implement these functions.

Design Methodology For GaAs Components

Three generic design philosophies are available for the development of high performance GaAs integrated circuits. The first and best known of these is custom chip design. Although the greatest on-chip performance can be achieved thereby, custom design may be unacceptable for GaAs integrated circuits in a rapid-turnaround environment, in which schedule and cost are as critical issues as is performance. Design costs in both manpower and computer time are too high, and the design cycles too long, for integrated circuits likely to be manufactured in relatively small quantities in the near and intermediate term.

A second approach which clearly circumvents the time consuming and costly custom design cycle is the use of configurable gate arrays or configurable cell arrays. Because the total "customization" required for gate or cell arrays is two (or occasionally three) metal layers and one via layer, chip turnaround from initial design to available parts can be a matter of weeks (provided that the appropriate computer aided engineering software packages are available). Considerable development in this regard for several GaAs structures is presently under way (1). However, the rather coarse spacing and less efficient packing density of the active elements in a configurable cell array (Figure 2) results in loss of on-chip speed performance, which can be tolerated in, for example, low power, radiation hardened environments but not in maximum performance processors.

The third possibility for semi-custom chip design, and the best way to achieve maximum speed performance short of a custom chip design, is the use of so-called "standard cells". In the standard cell approach, large functional macros of 50-200 gate complexity are custom designed to the transistor level and heavily simulated to verify and guarantee their performance. The macros are carefully selected for their wide applicability to numerous signal processing problems and their ability to be configured in a variety of ways as building blocks to assemble much more complex functions. The standard cells are essentially custom designs at all mask levels, but because their layouts remain constant over time, the individual mask layers for each cell can be stored on a library and recalled for repeated use. A number of standard cells are laid out on the chip, with "customized" interconnects designed to coordinate the standard cells into a complete functional entity. Since logical and analog simulations can be performed with extreme rigor on the standard cells, ultimately this approach provides a very low risk, rapid turnaround method of achieving integrated circuits with high levels of performance.

There are, however, a number of caveats to the use of standard cells in GaAs technology at present. First, Gallium Arsenide substrate and device fabrication technologies are evolving rapidly; the substrate and transistor characterization data which must be available to allow detailed simulations of standard cells is either lacking or in a state of flux. (Much less detailed simulations are required for equivalent levels of confidence in a design based on a configurable gate or cell array.) Second, there is as yet no base of experience in using rapid turnaround computer aided engineering software for digital Gallium Arsenide devices (e.g., for gate and cell arrays). Third, the use of standard cells in Gallium Arsenide has not yet been validated, particularly given the large number of Gallium Arsenide transistor and gate types currently under development. It is thus a somewhat high risk effort to develop a comprehensive standard cell library of perhaps 30-100 cells at present, since technology evolution might require a complete redesign and simulation of the library elements, perhaps several times. However, because of the long term efficacy of the standard cell approach for high performance GaAs components, a cost effective approach should be identified to allow investigation of this methodology in the near term.

The best approach to the CAE software to support standard cells must also be investigated. Standard cells can either be of completely variable rectangular size and aspect ratio, as has come to be the case for the software under development for military VLSI; conversely, so-called "standard height" cells can be employed. In the latter approach, cell widths can vary considerably, but a fixed vertical dimension for the cell is established which allows the individual cells to be positioned in rows (or columns), with the regions between the cell rows used for routing channels. In fact, this was the original meaning of the term "standard cell". The recent experimentation with variable aspect ratio cells in some of the military VLSI software developments have caused some confusion of terminology. Although not as general or flexible as standard cells of random aspect ratios, standard-height cell concepts are far better developed and their use will demonstrate nearly all of the features of variable aspect ratio standard cell design of integrated circuits, without placing nearly as much burden on the placement and routing software. Experience gained by the investigation of both of these approaches will be used during the next 3-5 years to select the most optimum mechanisms for implementing GaAs chips in standard cells.

ACKNOWLEDGEMENTS

Mrs. E. Doherty for preparation of text and Mr. S. Richardson for preparation of figures. This research was supported in part by contracts F33615-79-C-1875 from the U.S. Air Force, N00014-81-C-2661 from the U.S. Navy, and MDA903-82-C-0175 from the Defense Advanced Research Projects Agency.

REFERENCES

1. Yuan, H., 1982, "GaAs bipolar gate array technology", proceedings of the IEEE 1982 Gallium Arsenide Integrated Circuit Symposium IEEE #82CH1764-0, pp 100-103.

2. Lincoln, N., 1982, "Technology and design tradeoffs in the creation of a modern supercomputer", IEEE Trans. on Computers, C-31(5), 349-362 (May).

3. Gilbert, B. K., T. M. Kinter, S. M. Hartley, and A. Firstenberg, 1983, "Exploitation of Gallium Arsenide

Digital Integrated Circuits in Wideband Signal Processing Environments", proceedings of the 1983 IEEE Gallium Arsenide Integrated Circuits Symposium, October 25-27, Phoenix, Arizona. (In Press)

4. Gilbert, B. K., S. K. Kenue, R. A. Robb, A. Chu, A. H. Lent, and E. E. Swartzlander, Jr., 1981, "Rapid execution of fan beam image reconstruction algorithms using efficient computational techniques and special-purpose processors. IEEE Trans. on Bio. Eng., BME-28(2):98-116 (February).

5. Swartzlander, Earl E., Jr., Barry K. Gilbert, and Irving S. Reed, 1978, "Inner product computers", IEEE Trans. on Computers", C-27(1):21-31 (January).

6. Duff, M.J.B., 1982, "Parallel algorithms and their influence on the specification of application problems". In: Preston, K. and L. Uhr: Multicomputers and Image Processing, Algorithms and Programs. New York, Academic Press, Inc., pp 261-274.

7. Kung, H. T. and S. W. Song, "A systolic 2-D convolution chip". Ibid, pp 373-384.

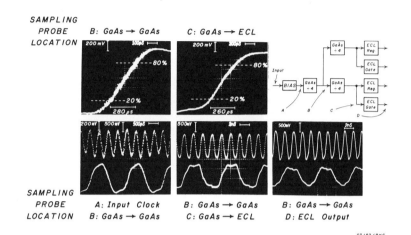

Figure 1 Operation of a small brassboard demonstration processor developed in the Mayo laboratories. Circuit board combines interconnected GaAs and silicon ECL components. Measurements are recorded from different locations in the circuit. System clock rate was 2.07 GHz.

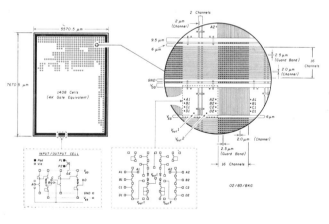

Figure 2 Logical and physical schematic layouts of a GaAs configurable cell array presently in design and test. Each internal cell is equivalent to several 3-input NOR gates. Some speed performance is wasted in a cell array implementation, but design time is minimized. Gate delays will be in the 100 psec range. (Reproduced with permission from Gilbert, et.al., "Exploitation of Parallelism and Ultraspeed Integrated Circuits in the Next Generation of Distributed Super Signal Processors, published in The Institute of Electrical and Electronics Engineers. (In Press)

SIMULATION FROM SYSTEM DESIGN TO CHIP DESIGN

K G Nichols

University of Southampton

INTRODUCTION

The design of a system on silicon usually proceeds through a number of largely distinct phases as shown in figure 1. Even without CAD facilities designs will be completed and chips will be fabricated. Such is the nature of human endeavour. Unfortunately, in many instances, the yield of circuits of acceptable performance will be low or zero because of design faults and very costly iterations around the design-fabrication loop will be inevitable.

Errors in functional design are common and very costly unless taken out of the major iterative loop. Validation of the functional design by interations around the minor loop using either formal verification techniques or simulations is a cost-effective approach to this problem.

Iterations around a logic design and simulation loop involve a trade off between pessimism and confidence because gate-delay characteris ation is inexact without consideration of layout. Further logic simulation using delays extracted from mask analysis is a necessary step in full validation of chip design, particularly in pursuit of high performance.

Similar comments apply to circuit design and simulation in that layout parasitic elements effect the characterisation of primitive circuits.

The computational cost of mask analysis, circuit-level simulation and, to a lesser extent, logic simulation at present forces modularisation in design, possibly at the expense of integration density and optimum performance.

An illustration of why it might be difficult to avoid circuit-level simulation based on layout data of quite large subsystems is given in figure 2. Until the layout design is complete it is not possible, for each gate, to evaluate the static and dynamic erosion of noise margins arising out of shifts of bus voltages. The shifts are caused by current flow through the parasitic resistances of the buses and by changes of charge distribution on the parasitic capacitance. Dynamic variations of noise margin manifest themselves as imprecisions of propagation delays and thus influence critical- path timing analyses. Any trend towards smaller feature size, higher degrees of integration, and reduction of wiring area, are matched by an increased significance of layout parasitics. The hierarchical design strategies, proposed for LSI and VLSI, will not necessarily avoid the need to verify the design against bus- shift voltages by analysis of the complete layout. This can be avoided only if conservative design constraints are accepted, or if some means can be found to effect characterisation of parasitics in a way that permits the characterisation to be carried up through the levels of hierarchy.

There is a general awareness of the use, advantages and disadvantages, of logic and circuit level simulators. In this contribution, therefore, attention is focussed primarily on functional simulation and macro-level circuit simulators.

FUNCTIONAL VALIDATION

At the outset of a design project there exists a concept of what the system being designed is required to do. The first step in the design is to capture this concept in a specification of the operational requirement of the system. This step is intuitive and cannot be automated.

As a trivial example, consider the specification of the two-dice game given in figure 3. An attempt has been made to write this specification systematically, even algorithmically. By an obvious translation step, this specification is equivalent to the state- transition graph shown also in the figure. Nevertheless, there is no guarantee that the concept of the game has been correctly captured. Simulation can improve confidence in the specification but cannot prove its correctness. In particular, simulation is limited to the input test sets used.

Manual translation of the specification into a hardware structure, e.g. a 1-bit register to represent the two states and control logic to represent the edge transitions, is another source of potential error which simulation might detect. Now the simulator must specifically model the logic of the hardware if not its physical realisation, If, however, the language specification could be automatically translated into a hardware structure, this second source of error can be eliminated. But then, the specification language must be precisely defined through its syntax which also makes possible automatic simulation of systems defined by statements written in the language. A dichotomy arises because a very high-level context dependent language with a rich syntax is best for capturing concepts but this makes the system description difficult to interpret or compile into a hardware structure and to simulate. The closer the language to hardware, the simpler the translation but the greater the difficulty in capturing the concept correctly. This is illustrated in figure 4 which lists a program in the SCHOLAR language (1) describing a 'hardware' bit-reverser. I defy anyone to ensure that the concept is correctly captured here without at least some manual simulation!

Two problems with 'batch' simulation in the above context are the determination of an appropriate input test set (often very large even for small systems) and the interpretation of the vast set of simulation output data. Both sets are important relative to confidence in the design. The same problems are pertinent also to logic simulation. Complete simulation of the bit-reverser above for 100% confidence level requires 2**16 input words! While the foregoing is a simple example, the problem of checking out, say, a floating-point hardware circuit is virtually insoluble. This approach, however, fails to capitalise on the intelligence of the design or of the designer.

In one address (2) to these problems, additional concepts are introduced. First, an interactive simulator similar in principle to the debuggers used in microprocessor development systems. This provides the designer with commands to control the simulation, e.g. code breakpoints and traps, sensing and setting of variables, traces, etc. Such a simulator would commonly be used in initial investigation of a new design to detect, in particular, the more obvious faults. Second, a batch simulator. This is essentially similar to the interactive simulator, but is invoked as required by a control program. Third, a set of primitives which form, in effect, a language for writing control programs. The objective is to specify the simulation environment for a SCHOLAR program in high-level language statements. Powerful control structures, output formatting, and input/output collating primitives are provided which allow the designer to specify intelligent tests for correct performance and to control the bulk of simulation output data.

Over the last decade considerable progress has been made in formal verification of programs. The techniques developed can be expected to find increasing applications to validation of systems defined in high-level languages. They are based primarily on algebraic modelling of programs and, in principle, the most powerful technique is algebraic reduction of a program. This permits the transformation between input and output to be verified against the required specification. As yet, algebraic reduction can only be effected for simple control structures and recourse to partial verification aids is the most that can be achieved. An example is the SPADE (3) suite of aids. This comprises a control-flow analyser, a data-flow analyser and an information-flow analyser. The aids can be used to detect anomalies in control structure, undefined and unused variables, unrecognised constraints and invariants, and information propagation paths. As such, the analysers are very useful tools in the production of semantically correct programs.

LIMITATIONS OF CIRCUIT-LEVEL SIMULATIONS

The computational cost of time-domain simulation of large nonlinear circuits is prohibitive. The cost lies principally in function evaluations for the Jacobian but also the whole of the Jacobian (coefficient matrix) must be resident in memory if equation solution efficiency is not to be impaired. Despite the falling cost of computer memory, this remains a problem because of demands in LSI and VLSI to analyse larger and larger circuits (e.g. for the bus-shift problem). The present practical limit is, for this, probably a circuit of about 1,000 nodes but computation times of hours and memory usage of megabytes would be needed.

Simulation algorithms incorporating network tearing strategies (4) can alleviate the memory problems but not the computation time problem even when latency is exploited.

MACROMODELLING

The need for significantly lower cost simulations than those obtainable with circuit analysers has caused macromodelling to become a widely researched topic (5). At one extreme, any model simplification technique, e.g. table look-up modelling, is a form of macromodelling. Such techniques are frequently used for operational amplifiers (6) and for logic circuits (7). Even the network tearing method, in which a subcircuit is reduced to a simpler equivalent circuit (8), is a type of macromodelling. At another extreme, functional macromodelling is used (9). Here the macromodel is characterised through its input/output relationships without regard to the physical interior of the subcircuit being modelled. This last form of macromodelling can reduce simulation times, in comparison to circuit level simulation, by factors of over a thousand. Figures 5 and 6 show a digital system and waveforms obtained using such a simulation. Computation time was 43 seconds and memory usage has then 64 kbytes on a large powerful minicomputer.

The principal use of simulation in digital-system analysis is to detect timing and race hazards which might cause the circuit to malfunction catastrophically. Part of the problem is determining which signal paths in the system are critical with respect to timing. Traditionally, a logic simulator would be used for this purpose with an assignment of a time delay to each gate in the system. Such inertial delays represent very crude modelling of the propagation characteristics of the gate and, accordingly, worst-case or confidence-interval minimum and maximum delays would be assigned. It is known that these practices give rise to very pessimistic assessment of satisfactory operation, particularly for large systems and where long logic chains occur.

The problem is that any macromodelling approximations to the characteristics of a gate compromises the precision of the very quantity which is needed most accurately, namely, propagation delay. The dilemma this causes is such that there is a considerable body of opinion that nothing short of full circuit-level simulation, with accurate modelling of all circuit elements, will determine critical timing paths in competitively-designed VLSI circuit chips. This view has stimulated research into accurate characterisation of macromodels (10) in pursuit of cost-effective simulation of large systems.

THE FUTURE
- HIERARCHICAL HYBRID SIMULATORS

The way forward in the development of time-domain simulators for support of VLSI chip design will almost certainly be based on hierarchical network-tearing strategies wherein each subcircuit is analysed separately. Not all subcircuits will, however, necessarily have to be analysed with equal precision. It is possible that some could be macrosimulated, some logic-timing simulated, and some even hybrid simulated. The concept of hybrid simulation is extended to imply that different subcircuits, at any level in the hierarchy, could be analysed by different types of simulator. Such general purpose hybrid hierarchical simulators are currently under development (11).

REFERENCES

1. Allerton, D. J., Batt, D. A., and Currie, A. J., 1983, Private communication, Electronics Department, University of Southampton.

2. Ibid.

3. Carre, B. A., 1983, Private communication, Electronics Department, University of Southampton.

4. Nichols, K. G., 1982, Proc. CAD82 (Brighton) 227-230.

5. Hsieh, H. Y., Rabbat, N. B., and Puebli, A. E., 1979 Proc. IEEE ISCAS, 336-339.

6. Glesner, M., and Wisang, C., 1975, Proc. IEEE ISCAS, 225-258.

7. Chawla, B. R., and Gummel, H. K., 1975, IEEE Trans CAS22, 901-910.

8. Linardis, P., and Nichols, K. G., 1979, Prox. IEE CADMECCS, 105-109.

9. Da Costa, E., and Nichols, K. G., 1980, Proc. IEE, Part G, 127, 302-308.

10. Fyson, C. J. R., and Nichols, K. G., 1982, Proc. IEEE ICCC, 189-192.

11. Zwolinski, M., 1983, Private communication, Electronics Department, University of Southampton.

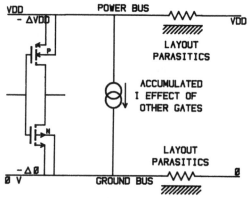

FIGURE 2. EXAMPLE OF BUS-SHIFT PROBLEM

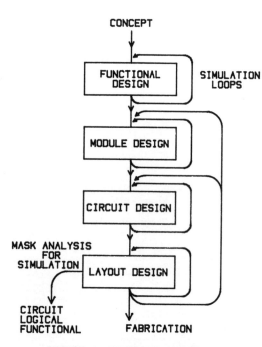

FIGURE 1. PHASES OF DESIGN

TWO-DICE GAME - RULES

GAME: THROW DICE
      IF 7 OR 11 THEN WIN & GAME
                     ELSE SAVE SCORE

RETRY: THROW DICE
      IF 7 OR 11 THEN LOSE & GAME
      IF EQSCORE THEN WIN & GAME
      RETRY

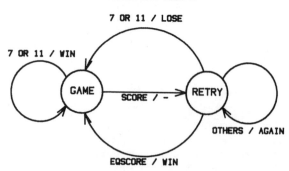

FIGURE 3. TWO-DICE GAME EXAMPLE

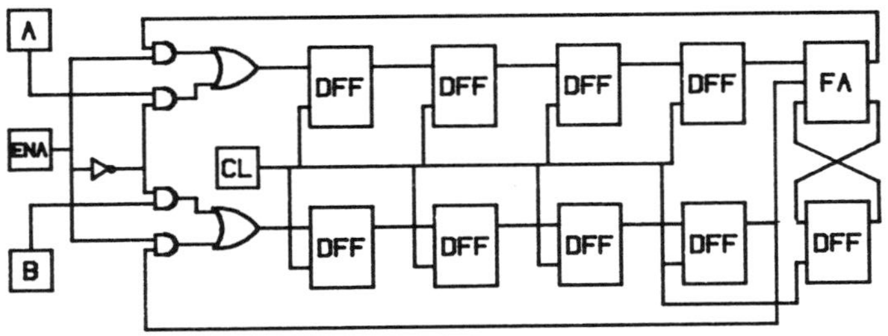

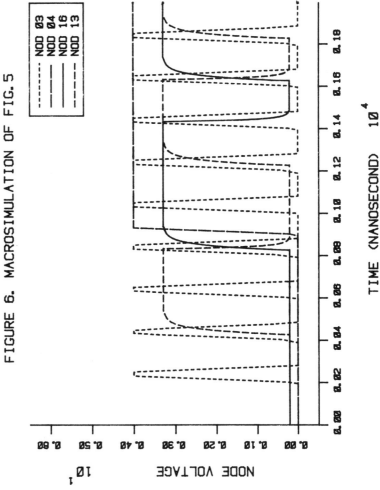

FIGURE 5. FOUR-BIT SERIAL INPUT ADDER

FIGURE 6. MACROSIMULATION OF FIG. 5

```
1   PROGRAM BITREV(WORD, IP.READY/
2                  WORD, IP.READY, OP.READY)
3   $(
4       DEFINE
5       $( WORD<15>
6          IP.READY,     OP.READY
7          TEMP<15>,     M<15>
8          MASK<15>  ROM<0:7>
9              #X8001,#X4002,#X2004,#X1008,
10             #X0810,#X0420,#X0240,#X0180
11      $)
12      OP.READY := FALSE
13      $( WHEN IP.READY DO
14         $( IP.READY, OP.READY := FALSE
15            FOR I = 0 TO 7 DO
16            $( M := MASK[I]
17               TEMP := WORD & M
18               UNLESS TEMP = 0 ! TEMP = M
19                  DO WORD := WORD XOR M
20            $)
21            OP.READY := TRUE
22         $)
23      $) REPEAT
24  $)
```

FIGURE 4. SCHOLAR PROGRAM

HIERARCHICAL DATA CAPTURE : LINKING SYSTEMS AND CHIPS

P. Blackledge

GEC Telecommunications Ltd., U.K.

INTRODUCTION

The technology of VLSI provides both enormous freedom and enormous problems. It permits the design of complete systems on a chip but, if the system becomes more complex than "one headful", the two-dimensional nature of layout on silicon places significant constraints on any division of labour between designers. The high setup costs to produce prototypes, together with the impossibility of "cut and hack" to fix design errors, result in an environment where there are high cost benefits in getting the design right first time.

We therefore require some means of organising the design task so that it can be divided between designers or done by one designer as a sequence of smaller tasks. Of these two, it is the case of multiple designers which creates the larger problems, as this involves the communication of information about the design tasks between people. In this paper we examine how design languages are relevant to this problem, and discuss the applicability of various types of design language.

HIERARCHY IN SYSTEM DESIGN

Whenever a design task is split between designers some hierarchical structure has been introduced, even if the system design is not being done in a "top-down" manner. The system has been split into sub-systems, and these sub-systems may perhaps be split into sub-sub-systems, etc., until the stage is reached where the units to be designed are each "one headful". The difficulty which this sub-division creates is that of ensuring that each designer is fully aware of the task to be performed, so that the sub-system designs do fit together and do interwork as intended.

The system design must therefore be documented in a way (see Figure 1) which provides precise and complete descriptions of:
(a) the required behaviour and performance of the system (sub-system, etc.),
(b) the split into sub-systems (sub-sub-systems, etc.), including all the interfaces between them,
(c) the required behaviour and performance of each of the sub-systems (sub-sub-systems, etc.).

NOTATION IN DESIGN

In the past, hardware and computer designers have found it necessary to use various forms of notation to describe design tasks and designs. Stylised graphical representations (e.g. circuit diagrams) and mathematical expressions (e.g. Boolean algebra) have provided a level of precision and comprehensibility which is almost impossible to achieve in English (Japanese, French, etc.). The restricted form of the notations has also made it possible to provide computer assistance in creating, editing and checking the design descriptions. This is even more significant when we are concerned with describing complex systems. The normal "clerical" error rate of humans is a significant problem which makes computer-based checking a necessity.

We must recognise that any description of a design is an abstraction; it omits much information in order to concentrate on particular aspects. For example, in the area of hardware design this has led to the creation of a number of languages at different abstraction levels (e.g. SPICE (1), TEGAS (2), ISPS (3)). Although there are advantages in having a separate language for each level, in that their syntaxes can be optimised to the task in hand, it does have the disadvantage that designers must learn and use a multiplicity of languages. The prospects for a single language to cover all levels of the design are discussed later.

As a generalisation, notations (languages) fall into two distinct categories:
(a) structural descriptions (e.g. circuit diagrams), which show how a system (sub-system, etc.) is constructed from simpler elements but do not contain any explicit information about the behaviour of the system,
(b) behavioural descriptions (e.g. Boolean algebra), which state what the system must do but give no information about its construction.

These categories match the two parts of the system documentation as depicted in Figure 1, and we require a language which permits both structural and behavioural description.

SPECIFICATION VERSUS DESCRIPTION

At this point it is necessary to make a slight diversion from the main topic of the paper in order to clarify the terminology being used. There is a wide range of terms to describe languages used in the design process, of which the following are a sample:
   specification language;
   design language;
   behavioural description language;
   hardware description language;
   circuit description language.

Philosophical arguments to decide whether X is or is not a "specification language" are irrelevant here, as are the distinctions (if any) between the various types.

The single term "design language" is used in the following sections to mean a single language which can be used to describe both structural and behavioural information. A design language is therefore capable of expressing both a purely behavioural descrip-

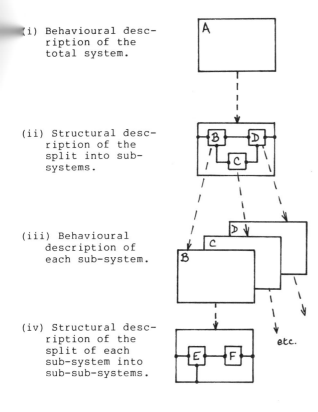

(i) Behavioural description of the total system.

(ii) Structural description of the split into sub-systems.

(iii) Behavioural description of each sub-system.

(iv) Structural description of the split of each sub-system into sub-sub-systems.

Figure 1 The Design Documentation Hierarchy

tion (e.g. a specification of a complete system) and a purely structural description (e.g. a circuit diagram). Both types of description are necessary in the hierarchy, and in the following sections we will assume that both are covered by a single language which we shall call DL.

FEATURES OF A DESIGN LANGUAGE

The best way of determining what features a design language should posses is to examine the way in which it is to be used. In what follows we will assume that the DL is to be supported by a wide range of computer-based tools for editing, checking, etc..

Let us look at a system which is being designed and completely documented in DL. Starting from the top levels of the design hierarchy (see Figure 1), the sequence of events should be as follows:
(a) produce a DL description of the required behaviour and performance of the complete system,
(b) confirm that this DL description correctly represents the required behaviour,
(c) make some design decisions which split the system into sub-systems,
(d) document in DL both the structural description of the split into sub-systems and the required behaviour and performance of each of the sub-systems,
(e) confirm that the sub-systems together (as defined in their DL descriptions) produce the same overall behaviour as the description produced in (a),
and so on for the design of sub-sub-systems,

etc.. The important stages are (b) and (e), where checking activity takes place. This checking should trap errors early in the design process, and it is this which produces the cost savings to justify the extra effort involved in producing DL descriptions.

At the level of the complete system (activity (b) above), the checking can be achieved by design reviews of the DL documents, or possibly by using the DL descriptions as input to a simulation system. Whichever method is used, the DL description can only be compared with ideas in someone's head or with some English text; there is no way in which human judgement can be replaced at this stage. The use of simulation seems essential here for complex systems, as the implications of long DL descriptions may be difficult to deduce simply by reading them.

For the checking at other levels (activity (e) above) the situation is different, in that there are two DL descriptions to be compared. Here the use of human reviewers to compare two large descriptions is likely to be unsatisfactory, so that we must look for other methods. There are two possibilities: simulation and formal proof.

Simulation with two descriptions involves the provision of the same inputs to two simulation runs, one using the system level DL description and the second using the interconnection of the sub-system DL descriptions. The output from the two runs have to be compared to check that they are identical; this is a job which can be done by a computer program. In general, simulation has the advantage that it is a familiar and widely-used method; its disadvantage is that the results are only as good as the set of inputs. A poor set of inputs will only exercise part of the description, leaving the other parts unchecked.

Methods of formal proof of correctness are based upon theorem proving in pure mathematics, and rely on the design language having a well-defined (i.e. mathematically sound) syntax and semantics. They involve providing a mathematical demonstration (a proof) that the two DL descriptions (the system description, and the interconnection of the sub-system descriptions) are equivalent. This therefore produces a more comprehensive check than simulation, but is a relatively new technique which has not yet been demonstrated on large and complex systems. Another difficulty at present is that computer support environments for the proof process require large amounts of processing power, and even then may not be able to handle large and complex descriptions.

By examining these possible uses of DL descriptions we can extract a number of features which DL should possess. These are:
(i) well-defined syntax and semantics, so that the computer-based support tools can be correctly constructed,
(ii) a syntax which eases the job of writing and reading descriptions of large systems,
(iii) a means of interpreting DL descrip-

tions as code for use in a simulation system,
(iv) if DL is to be used in proofs of corrrectness, a relatively simple syntax and semantics,
(v) representations for time and other performance factors.

Additionally, recognising that DL is intended to tackle "real world" problems, it must be possible to work with, and check, incomplete descriptions, to allow the documentation to be prepared as the design project progresses.

A BRIEF REVIEW OF DESIGN LANGUAGES

The number of existing languages runs into hundreds, so that it is not possible to give a comprehensive review here. Instead the main groups of languages are discussed, and a small number of each type mentioned by name. This makes the review relatively superficial, and the named languages have been selected on a highly subjective basis. Groups of languages have been omitted because they do not seem to come near to being well-defined, but it must also be admitted that some of the languages mentioned are poorly defined. Longer and more specific reviews of design languages can be found in (4) and (5).

Certain of the features mentioned above are not satisfactorily covered by any type of language, these being time, performance factors, and the ability to work with incomplete descriptions. Time and performance factors are a very complex area, in that it is not clear what sort of information is appropriate and where it should appear in descriptions, but this represents a major failing in all the languages. The inability to work with incomplete descriptions is probably more a feature of the computer support tools associated with a language than the language itself, and is therefore not as serious.

Hardware Description Languages (HDLs)

Historically, HDLs have progressed bottom-up, starting from gate level descriptions (e.g. TEGAS (2)) and moving up to register transfer level (e.g. AHPL (6) and ISPS (3)). This is reflected in their concentration on structural description and their relative weakness in the behavioural area. They are also limited to the level of individual wires and bits (or at best, vectors of wires and bits), so that they cannot be used at higher levels of system description where signals are often best represented as characters or numbers or just messages.

HDLs posses feature (iii) of those listed in the previous section (ability to be simulated), but are not easy to write and read (feature (ii)). Although the syntax of these languages is usually well-defined, the semantics are only stated informally. This may be corrected in some new HDLs, such as the US DoD VHDL project (7).

Simulation Languages

This group includes languages specifically designed for use in hardware design (e.g. ELLA (8)) aswell as general purpose simulation languages (e.g. GPSS (9) and SIMULA (10)). They are capable of describing behaviour and structure at all levels of the design hierarchy, are definitely suitable for simulation (feature (iii)), and have well-defined syntaxes (part of feature (i)).

With only a few notable exceptions (such as ELLA) they are similar to standard computer programming languages; this has generally been done in the interests of run-time efficiency of the simulator but has two disadvantages. The first is that the semantics of this type of language are complex and make them unsuitable for use in formal proofs, whilst the second is that descriptions written in these languages are often unclear and difficult to read (feature (ii)).

The penalty paid for the ability of these languages to cover all levels of the hierarchy is a considerable loss in simulation efficiency when compared to purpose-built simulators for any single level (e.g. logic gate level). However, if the design is done in a proper hierarchical manner then this may not be significant, especially in view of the advantages of having a single language to cover all levels.

Formal Description Languages

Languages such as CCS (11) and META IV (12) are capable of describing systems behaviour at all levels. Some (such as CCS) also cover the structural description, although most don't. All the languages of this type tend to produce very clear descriptions of behaviour, as they provide ways of stating the required results independantly of any method for achieving these. Unfortunately, this clarity is often obscured by their unfamiliar notation, together with the lack of suitable introductory texts; this is an education problem which has not yet been tackled satisfactorily.

By their very nature these languages do have well-defined and simple syntaxes and semantics (features (i) and (iv)). They do, however, tend to permit the description of behaviour in a way which makes simulation difficult; this is a direct consequence of allowing descriptions in terms of the required results rather than in terms of a method for achieving those results. It is also not yet clear if these languages retain their clarity of expression when used on large and complex systems, as no large examples seem to exist.

SUMMARY

As can be seen from the above review, there is no existing ideal design language, and there may never be because the features required in an ideal language are to some extent contradictory. The pressing need for such a language in practice means that organisations will have to make some sub-optimal decision, and for this reason the author offers the following subjective apraisal.

Formal languages and the methods of proving correctness, although they hold great promise for the future, have not yet been demonstrated on large systems designs. They are therefore not a practical choice for commercial organisations at present. A number of UK organisations are involved in trials of a variety of such languages, and these trials should produce an indication of the viability of the approach. This type of trial is also valuable in providing experience of the mathematics and proof methods which is necessary to evaluate new languages and ideas as these appear.

As HDLs do not cover all levels, the short-term choice therefore appears to lie between using a single simulation language and using a mixture of languages (e.g. a simulation language for the higher levels and an HDL at lower levels, etc.). A single language is much preferable, in that it reduces the amount of staff training required and makes checking of adjacent levels easier. The run-time simulation efficiency at logic gate level should not matter, as it should not be necessary to do much simulation of the complete design at this level.

Of the simulation languages, ELLA (or any other similar product) looks a good choice. It is purpose-designed for the description of hardware, but is not limited to bit level. It has a simple syntax, and the computer support system performs many checks on the descriptions (e.g. to detect inconsistent interpretation of messages by the sender and receiver). As an addditional benefit, it is a simple language which bears a strong resemblance to some of the formal behavioural languages so that its use may prepare the way for the later adoption of formal methods.

REFERENCES

1. L.W. Nagel, "SPICE2: A Computer Program to Simulate Semiconductor Circuits", Memo ERL-M520, University of California, Berkeley, May 1975.
2. S.A. Szygenda, "TEGAS 2 - Anatomy of a General Purpose Test Generation and Simulation System for Digital Logic", Proc. 9th Design Automation Workshop, June 1972.
3. G. Bell and A. Newell, "Computer Structures: Readings and Examples", Digital Press, Maynard, MA, 1973.
4. W.M. vanCleemput (ed.), "Tutorial: Computer-Aided Design Tools for Digital Systems", IEEE Computer Society, 1979.
5. P. Blackledge, "The Selection of a Specification Language", IEE Conf. Pub. 198, 1981, 25-30.
6. F.J. Hill and G.R Peterson, "Digital Systems: Hardware Organisation and Design", Wiley, NY, 1973.
7. G.W. Preston (ed.), "Report of the IDA Summer School on Hardware Description Languages", DTIC ref AD-A-110866, Oct. 1981.
8. J.D.Morison, N.E. Peeling and T. Thorp, "ELLA: A Logic Language and Simulator", RSRE, England, July 1981.
9. -, "General Purpose Systems Simulator III", Form B20-0001, IBM Corp., White Plains, NY, 1963.
10. O-J. Dahl and K. Nygaard, "SIMULA: An Algol-based Simulation Language", CACM, 9(6), 1966, pp 671-8.
11. R. Milner, "A Calculus of Communicating Systems", Lecture Notes in Computer Science No. 92, Springer-Verlag, Berlin, 1980.
12. C.B. Jones, "Software Development - A Rigorous Approach", Prentice-Hall, London, 1980.

CHAIRMAN'S REPORT - SESSION 2 "CAD TOOLS"

G.F. Vanstone

Racal Microelectronic Systems Limited, UK

Within the five papers permitted in this session, it was only possible to select key areas within the spectrum of CAD aids that have to be available to the VLSI systems designer. The papers covered the topics of Description Languages and their role, Database organisation, Automatic Placement and Routing, Testability Analysis and Testing Methodology.

A VLSI design system will have to cope with many descriptions or databases associated with a particular design, the system and circuit description, simulation, layout, test description etc. - the organsiation and communciation associated with these descriptions is crucial to obtaining an efficient VLSI design environment. Russell et al reviewed some of the basic requirements for such a system with particular emphasis on a powerful hierarchical structure and the use of a central database as approved to a file based system. By describing the history of database development within IBM, the authors showed how their experience in these developments have led them to propose a new form of database organisation, the entity set/Binary relational database. The essential simplicity and elegance of their approach is attractive and it should provide good communication between design levels and descriptions. Hardware limitations to date have limited the full application of the database approach to VLSI design but the authors were confident that their future investigations would demonstrate its viability.

It is still generally accepted that in the layout of integrated circuits that the machine has a long way to go to equal the efficiency in functions per unit silicon area achieved by a good layout design, in some cases an automatic:manual ratio of 2:1 is quoted. However, good layout designers are scarce and manual layout is expensive, time consuming and open to error. For these reasons the use of CAD tools to automatically layout circuits is beocming predominat. Yanagawa pointed out that great progress had been made with currently available placement and routing techniques and illustrated this with the automatic layout of a 20,000 gate array design and 32-bit microprocessor chip which had been laid out automatically. The author highlighted the fact that it was in the placement area that most improvement appears possible. Particularly, he mentioned the intelligent automatic placement of cells taking into account the performance requirements of the chip.

In several papers of the seminar it was emphasised that a more formalised specification for describing integrated circuits becomes essential at VLSI levels of complexity. The use of full formal verification of logic with specially derived languages as a standard approach still seems to be some way off in the future. However, it was the contention of Thorp et al in their paper that existing hardware description languages such as ELLA could be improved to act as an interim form of specification language. The author illustrated this by showing several examples of the flexibility of the HDL in logic description. Clearly this is an important area of work for the future and would greatly aid the interface between the systems and VLSI designer if such a formalised approach to chip description could be implemented.

One of the major areas of difficulty with both LSI and VLSI remains the effective implementation of an overall testing approach. Keiner in his paper on testability analysis reviewed the various software techniques that have been used in the recent future to analyse the testability of logic designs. Many of these revolve around the use of sophisticated simulation programmes with some form of stuck-at fault analysis. In particular, the author was looking at the problem from the view point of a procurement agency dealing with design contractors where the agency have little or no control over the way detail design is carried out. To bring an element of standardisation into the procurement process, the author explored the definition of a testability figure-of-merit and also the specification of a new hierarchical test and simulation standard being generated by the US Dept. Defence. Whilst clearly important from the customer point of view, this attempt to impose a defined form of standardisation was criticised by other delegates as being too bureaucratic and over restrictive on design suppliers.

The final paper in this session was given by Dr. Williams from IBM who reviewed in detail the approaches for VLSI testing both at a chip level and at a board level. Again the vendor-customer interface problem was raised by pointing out that the customers lack of logic models made it extremely difficult for him to adequately test devices. For designers, the development of a testing methodology is vital to include such features as Self-Test Logic by, for example, signature analysis, the provision for on-going maintenance checks of an operating chip and designing with structured logic to ease the external testing problem.

DESIGN SYSTEM ARCHITECTURE AND THE ROLE OF DATABASE

P.J.Russell, N.Winterbottom, S.Newberry, J.M. Snyder

IBM United Kingdom Laboratories Limited, Hursley Park, Winchester

ABSTRACT

This paper proposes a VLSI design system based on a central database and discusses some important aspects of such a system. Following a review of database technology, it is proposed that the entity set/binary relational architecture provides a suitable model for VLSI design data. An entity set/binary relational schema which allows hierarchical design data and its interconnections to be modelled is then developed.

INTRODUCTION

During the 1970's a programmable system for checking LSI chips was developed at IBM Hursley (Russell (1), Newberry and Russell (2)). During the same period, a general-purpose database system called NDB was developed at Hursley (Sharman and Winterbottom (3), Winterbottom and Sharman (4)). The experience gained with these tools has led us to define a new design system which deals with the problems brought on or aggravated by the advent of of VLSI technologies. These problems include data volume and complexity, lack of tool integration, and a diversity of design representations and methodologies.

A NEW DESIGN SYSTEM APPROACH

The primary requirement of any engineering design system should be the establishment of an environment that stimulates and supports the engineering endeavour. If a VLSI design system is to successfully fulfil this primary requirement, we feel that it must

- be based on a central database,
- support hierarchical design and analysis,
- support multiple design representations,
- be user programmable in a suitable high level language.

One or more of these requirements have been discussed by a number of authors, including Rosenberg (5), Katz (6), Haskin and Lorie (7), but to our knowledge no one has yet combined this set of requirements into a functional system for supporting VLSI design. Let us now consider the above system requirements in more detail.

Central Database

A system based on a formal database architecture will do a number of things to enhance the design of VLSI chips. A database will aid tool integration by providing a single repository for all design data and presenting a consistent data interface. The locking and audit facilities that are typically part of an authentic database will reduce or eliminate problems with data integrity that might arise because of the multi-designer, hierarchical design environment required by VLSI. The ability to support complex queries and assertions in an ad hoc fashion will enhance the creativity and spontaneity of a designer by allowing him to add or retrieve data in a fashion that might not have been foreseen.

The database schema, which defines a user's perception of data structure, is of great importance. A poor schema will adversely impact the ability of a design system to support hierarchical design and the multiple design representations discussed below. A schema which we think is suitable for integrated circuit design is proposed in a later section.

Hierarchical Design and Analysis

The term "hierarchical" implies that the design approach allows a design to be broken down into constituent pieces through any number of nesting levels. Each piece can be individually designed, simulated, checked and locked against future update. As low-level pieces are reused in higher levels of the design the only checking required is that these pieces have been used correctly in the context of the higher-level environment.

To fully support hierarchical processing, a system must allow a lower-level hierarchical element to be represented with the appropriate degree of abstraction when a higher-level element is being processed. It must also flag any changes made in a lower-level element and notify any higher-level elements containing this element that a change has taken place so that application software can deal with it appropriately.

Multiple Design Representations

At any particular level, a number of different equivalent representations exist to aid the design process. These might include high level behavioural logic descriptions, design cell "bounding-box" definitions, physical mask descriptions, electrical schematic descriptions, PLA definitions and so on. A system must be able to keep track of the correspondence between different representations and, as in the case of hierarchical processing, notify all representations of an element when a change occurs in any of them.

User Programmability

Programmable tools have been widely successful and indicate the desirability of a programmable tool base. It has been demonstrated that programmable editors such as the VM/SP/370 editor, XEDIT, are significant time and labour-saving tools. XEDIT presents most of its power to a program interface, thus allowing a user to apply the power of the editor's data manipulation with a simple programming language. The SL1 language (2) has shown the power of a user programmable approach in the area of IC design checking.

User programmability enhances system intelligence, since it allows the system to be programmed locally to accommodate changes in technology, available function and the user interface. This allows the intelligent designer to customise his own system rather than rely on the system developers to do it.

A REVIEW OF DATABASE TECHNOLOGY

A database is a carefully designed repository for complex data that can bring some order into the sharing of common data across different applications. In this section, we will review three database models, the hierarchical model, the n-ary relational model, and the less well known entity set/binary relational model.

Hierarchical Databases

In the hierarchical model, data is stored in a tree structure. The characteristics of systems implemented using this model include the following:

- Detailed data manipulation is moved from the application program to the database system,
- There is some separation of the logical problem of data structuring from the physical placement of the data on the storage medium,
- Reasonable performance can be expected, provided that the mode of access is a good match to the particular hierarchical view initially selected,
- Extensions to existing database schema can be difficult,
- Significant expertise is required to maintain performance in the face of changing data access patterns.

The fundamental drawback of this approach is that the data model is not canonical. The selection of one particular hierarchical view from the many possible hierarchical views of a set of data stamps the model with the mark of some particular way of accessing the data; accessing the data in different ways may then be complicated and inefficient.

N-ary Relational Databases

In the n-ary relational model, data is stored in a collection of tables, the contents of which must be in a normalised form and which are subject to a defined set of update and retrieval operations. Each table consists of a number of rows, each of which contains "n" fields. One or more of these fields are designated as "key" fields. (There is extensive literature available on this subject to which the reader is directed for a more detailed explanation.) Some characteristics of the n-ary relational approach are:

- The representation of data is more nearly independent of any particular view of the data than in the hierarchical model,
- The table operations force a level of consistency in the data,
- The nature of the model (tables) usually leads to implementations which are file or record-oriented with consequential performance biases which may not match the performance requirements of a particular application.

There are now a number of relational database systems in use for commercial applications. Many of these are built on top of a sequential or index-sequential file base and inherit the performance characteristics of those storage mechanisms. As a result, considerable expertise may be required to tune these systems for high performance in a particular application environment. Although the n-ary relational approach leads to a much cleaner logical database design than does the hierarchical model, there has been much discussion about enhancing this model to accommodate more sophisticated data constraints (Codd (8)).

Entity Set/Binary Relational Databases

The binary relational model has powers of expression comparable to those of of the n-ary relational model, but is conceptually simpler. The version of the binary relational model we will discuss below is constructed using just two notions: the "entity set" and the "binary relation." As an illustration of the application of the binary relational model, an extract from an administrative database will be used.

An entity set is a collection of elements of a single type. Associated with each element is a single data value (character string, numeric value, etc.) which may be null. An element in the database stands for a single thing (object, value, idea, etc.) in the real world which the database is intended to describe. In the example of Figure 1, entity sets are represented by boxes; the name in the box is the name of the set and indicates the type of entity in the set.

A binary relation is a set of connections between pairs of elements in two defined sets. In Figure 1, a relation is represented by a line between two boxes. Associated with each relation is a "mapping." This defines the number of connections allowed between the elements. The mapping values allowed are the four possible permutations of "one" and "many." These are denoted by combinations of "1" and "M" in Figure 1. For example the notation on the line between PERSON and ROOM should be read as:

- For each PERSON there is at most one ROOM,
- For each ROOM there may be any number of PERSONs.

If there is any doubt about the meaning of a particular line it must be labelled and an arrow placed on the line indicating the semantic sense of direction. MANAGER, JOIN, and LEAVE are examples of labelled relations in Figure 1.

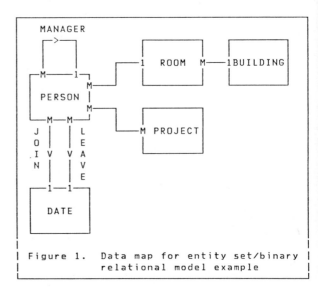

Figure 1. Data map for entity set/binary relational model example

The major advantages of the entity set/binary relational approach are its simplicity and a conceptual representation which is independent of possible machine implementations. The design of database schema is simplified to the point that verification of a design reduces to asking the following questions:

- Is it clear what kind of object the elements in a set stand for?
- Is the meaning of a particular binary relation well defined?
- For each element in a set S1, is there one or are there many elements in set S2 when we traverse a relation from S1 to S2?

Each question refers to an elementary component of the data map and the answer can be immediately compared to the facts of the real world for verification.

Since there are very few database systems in existence that are based on models with the inherent simplicity of the entity set/binary relational model, the benefits of such architectural simplicity have yet to be fully explored. It is already clear, however, that it allows a very simple interface to be used. It also appears that this simplicity makes it much easier to develop performance-oriented database software or to define simple algorithms for implementation in hardware.

A DATABASE MODEL FOR VLSI DESIGN SYSTEMS

The simplicity of the binary relational database model appears to make it attractive for a VLSI design environment in which design complexity is becoming a limiting factor in our ability to implement new systems. We must, however, ask two key questions about the suitability of this database architecture for VLSI design systems:

- Is it functionally rich enough?
- Can it be implemented with acceptable performance?

Over the period from 1975 to the present we have applied the binary relational architecture to a variety of data storage problems within IBM. Applications include numerous administrative systems, business chart specifications, engineering release control, economic and performance modelling, interactive drawing, system configuration specifications, and adaptive learning. Our experience indicates that the binary architecture leads to neat, understandable, stable, and expandable data models across a wide range of applications. We believe that it is functionally powerful enough to support complex VLSI data. It is also well suited to the implementation of a variety of intelligent interactive end-user languages.

Current systems based on the binary relational model do not have performance capable of handling the data volumes involved in VLSI design. However, since the development of such systems is still in its early stages, it seems reasonable to expect significant improvements in performance.

A SCHEMA TO MODEL STRUCTURED VLSI DESIGN

In this section, a simple latch macro (Figure 2) will be used as an example to show how binary relational techniques can be used to model hierarchically structured chip design data.

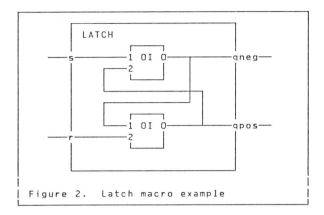

Figure 2. Latch macro example

Structure

We begin by defining a framework to model elements of the design and their use. In Figure 3, sets SPECIFICATION and INSTANCE are used to represent the definition and usage of these elements. Design hierarchy is modelled by the relations DEFINE and INVOKE. The sets and relationships are abstract and do not constrain the representation of the design data.

In the example, the "definition" of the OI and LATCH blocks is contained in the SPECIFICATION set. Each definition exists as an individual element in this set. Each "invocation" of OI in the LATCH block is modelled by a unique element in the INSTANCE set which has a connection to LATCH via the INVOKE relation. The detailed information particular to each INSTANCE, eg. block placement on the page, is associated with its corresponding element. Given one of the INSTANCEs of OI, its actual definition may be found by traversing the DEFINE relation to the element describing OI in the SPECIFICATION set. If LATCH were to be instantiated in a higher-level function, its definition would be found in a similar manner.

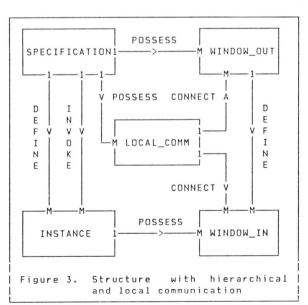

Figure 3. Structure with hierarchical and local communication

Hierarchical Communication

The schema can be extended to model communication between adjacent nesting levels. Two additional sets, WINDOW_OUT AND WINDOW_IN, are introduced to represent "communication windows" from a SPECIFICATION to the world outside and from the invocation of an INSTANCE to its inside world. The ownership of an element in these communication windows is modelled by a connection in the appropriate POSSESS relation. The DEFINE relation is needed to correlate these windows across hierarchy levels. This relation models the correspondence of a WINDOW_OUT element from a detailed design SPECIFICATION with the WINDOW_IN elements of its defined INSTANCEs.

In the example, each OI block INSTANCE has three communication windows to its inner definition (labelled 1,2 and 0 in Figure 3). These are represented by corresponding elements in the WINDOW_IN set. Connections in the DEFINE relation associate these WINDOW_IN elements with their original WINDOW_OUT definitions, which are in turn POSSESSed by the OI block definition in SPECIFICATION. In a similar fashion, the LATCH block has four windows defined to the outside world (s,r,qneg and qpos) that are modelled by elements in the WINDOW_OUT set. Note that the association between the original definition of a WINDOW_OUT element and its WINDOW_IN usages does not depend on any ordering or naming convention; it is explicitly stated by the DEFINE relation.

Local Communication

Another set, LOCAL_COMM, defines abstract connectivity detail within a SPECIFICATION. The ownership of an element in LOCAL_COMM by an element in SPECIFICATION is modelled by a connection in the POSSESS relation. The connection of elements in LOCAL_COMM with elements in the communication window sets discussed on the previous page is modelled via the CONNECT relations.

In the example, local communication elements model the nets which CONNECT to pins of the INVOKEd OI blocks (0, 1, 2) and to the pins of the LATCH macro being defined (s, r, qneg, qpos). The representational detail of a net, ie. its topology, is associated with the local communication element.

Communication Between Non-adjacent Levels

It is sometimes convenient to suppress communication through design levels. If purely hierarchical communication were used, two non-adjacent levels which need to communicate would force all intervening levels to be aware of it. This has the undesirable effect of increasing their complexity. However, non-hierarchical communication can be explicitly represented by the addition of a connection to a relationship which "globally connects" local communication POSSESSed by one SPECIFICATION with that POSSESSed by another.

A good example of this type of communication is the use of power and ground buses. These are important in the global chip context where their pad connections and current carrying capacity have to be considered, and in low level schematics where circuits have to be connected to ground and power rails. However in the intervening logic description levels, such as the latch macro example, they are irrelevant and unnecessary.

CONCLUSION

We have attempted to give the reader an overview of CAD system requirements for VLSI design. These have developed from our collective experience in the areas of CAD and database software, and are likely to evolve as we develop a better understanding of VLSI.

We have also described a database schema which provides a typeless skeleton for hierarchically structured design. The representational detail of different design domains can be associated at the nodes of this schema. It provides a non-ambiguous means of communication between design levels either within or across a hierarchy and can be used to focus interest on the current level by allowing other levels to remain hidden. The binary relational technique provides a simple, consistent idiom for both user and tool developer. The potential for high level language support to navigate design data in this form is good and will make application programming simpler.

A system based on the ideas presented in this paper is currently being investigated.

REFERENCES

1. Russell,P.J., 1974, "SIGMA," IBM Technical Report TR.12.121.

2. Newberry, S. and Russell, P.J., 1981, "A Programmable Checking Tool for LSI," European Conference on Electronic Design Automation, Brighton, England.

3. Sharman, G.H.C. and Winterbottom, N., 1978, "The Data Dictionary Facilities of NDB," Proceedings of the Very Large Data Bases Conference, Berlin, W.Germany.

4. Winterbottom, N. and Sharman, G.H.C., 1979, "NDB: Non-programmer Data Base Facility," IBM Technical Report TR.12.179.

5. Rosenberg, L.M., 1980, "The Evolution of Design Automation to Meet the Challenge of VLSI," Proceedings of the ACM Seventeenth DA Conference, Minneapolis MN, USA.

6. R.H. Katz, 1982, "A Database Approach for Managing VLSI Design Data," Proceedings of the ACM Nineteenth DA Conference, Las Vegas NV, USA.

7. Haskin, R.L. and Lorie, R.A., 1982, "On Extending the Functions of a Relational Database System," Proceedings of the 1982 ACM-SIGMOD International Conference on Management of Data.

8. Codd, E.F., 1979, "Extending the Data Base Relational Model to Capture More Meaning," IBM Research Report RJ2472(32359).

CHIP DESIGN USING AUTOMATIC PLACEMENT AND ROUTING

T. Yanagawa

NEC Corporation, Japan

INTRODUCTION

Layout design is the process of transforming a structural description of circuit into a physical pattern. This step is important because it determines chip area which has major influence upon chip costs, and thus the largest portion of design time is usually spent for this design phase in the case of manual design. Many reports on the development of automatic layout tools have been published since 1960s. At present this is the only design item where automating tools are widely utilized.

Automatic layout contributes not only to fast turnaround time but to error-free designs and low design costs. The present automatic techniques, however, cannot generally produce as high component density as skilled designers. Therefore, the optimum design strategy should be chosen according to the type of devices. Custom chips in contrast with standard chips should be designed automatically since they are produced in relatively small quantities and require fast turnaround time and low design costs. The trends to VLSIs will expand the market of custom devices and this in turn will enhance the importance of automatic layout techniques.

This paper reviews automatic placement and routing techniques for the two conventional types of custom LSIs, gate array and standard cell. Subsequently difficulties of the present techniques when circuit complexity increases further and some means to overcome them will be discussed.

GATE ARRAY

Layout Model

This approach is to complete layout only by designing interconnection patterns customly on an array of prefabricated cells. The design objects are (i) to select a master with as small number of unused cells as possible, and (ii) to achieve 100 percent routing.

Placement and Routing Methods

The selection of master is usually done manually. The placement and routing methods used in some recently published CAD systems are listed in TABLE 1. It is conceived that better results can be obtained by the combinational use of multiple methods rather than by a single method.

The placement is the key to 100 percent routing. The paper by Hanan et. al. describes experimental comparisons between placement methods (1). For large circuits, the initial placement is usually done by two steps, the group assignment and the block assignment. An object parameter for placement optimization is desirable to reflect upon the uniformity of net density distribution so as to avoid connection failures due to local congestion of nets.

The routing consists of two phases; the routing of as many nets as possible by highly efficient routers such as the line search router and the channel router, and the routing of remaining uncompleted nets by slower but more elaborate methods. The maze router is suited for the second phase since it always finds a path as long as it exists.

Discussions

As fast turnaround time is the subject of top priority for gate arrays, they should be designed automatically with least human interventions. However, successful routing cannot always be achieved unless excessively sacrificing component density. Therefore, it is unavoidable to prepare interactive methods to deal with uncompleted nets. Such systems should have intelligent functions rather than a simple pencil-and-paper function for efficient and error-free routings (2). The rip-up and rerouting methods have been developed as a means to automatically process uncompleted nets.

Routing areas on masters and cell utilization percentage are other factors which influence the probability of successful routing. Heller et. al. have statistically derived a relation between channel area and number of gates (3). The cell utilization factor of 90 percent is guaranteed for most commercial gate arrays.

Limitations

The up-to-date CAD systems for gate arrays can handle as many gates as 20,000. The mathematical computational complexity of typical placement and routing algorithms is shown in TABLE 2. The rule of thumb, however, indicates that the computer time is proportional to the 1.1-1.2 power of circuit scale, which is not so strongly nonlinear (4). The real limitation of circuit complexity for gate arrays, however, will stem from the restricted number of devices which can be realized by a simple array of gates.

STANDARD CELL

Layout Model

This approach is to represent a given circuit by standard cells which are predefined and stored in a cell library, and to layout these cells so as to minimize chip area. The fundamental rules are as follows:
1) Rectangular cells in a row have the same height.
2) Routing channels between cell rows are rectangular without any obstructions in them.

Placement and Routing Methods

The placement and routing methods used in recent CAD systems are listed in TABLE 3. The placement methods are essentially the same as those used in gate arrays. Proper arrangements of feed-through cells are important for the reduction of net length and chip area.

The main router for standard cell is the channel router because of variable channel width. Many improvements have been done on this router since its first development by Hashimoto and Stevens in 1971 (5). They are to attain routings even when cyclic constraints exist and to minimize channel width. Although being heuristic, these methods can realize channel width sufficiently close to the mathematical minimum. The application of channel router accompanies inevitably the preprocess step of determining a routing path of each net. This step is called loose routing which mostly uses the maze technique.

Discussions

Component density of standard cell LSIs is reported to be higher by 30-50 percent than that of their counterparts by gate arrays. Chip areas by automatic techniques, however, are sometimes twice as large as those carefully designed by hand. Therefore, the largest problem of this approach is to improve component density.

A typical phenomenon which wastes routing area is a congestion of nets in the middle of channels. One remedy is the use of loose routers with net density as object parameter. Considering that a fundamental difference of automatic techniques from manual designs is a lack of close interactions between placement and routing, another method will be the effective usage of interactive tools. Placement modifications based upon monitoring diagrams of net density distribution on CRT displays seems to be useful.

Another approach is the efficient usage of channel areas by applying meticulous design rules. The grid-free router by Sato et. al. eliminates waste areas caused by the fixed grid (6). Reduction of channel width by about 12 percent has been reported by this method. The application of clean-up process to reduce vias and bendings after detailed routing is another approach. The development of efficient routers for diagonal lines will be effective for compact layout.

Limitations

The CMOS 32-bit microprocessor chip with 78K transistors has been successfully designed by automatic layout technique (7). The computational complexity of placement and routing is empirically of the order of $O(n^{1-2})$ and $O(n)$, respectively, where n is the number of cells. The adoption of hierarchical concept is the effective means to cope with complexity problems. Component density decreases with the increase of n since channel area is proportional to $n^{1.5}$ (8). Like gate arrays, what limits the maximum circuit complexity of standard cell is the scarcity of circuits realizable by the simple layout model.

LAYOUT TECHNIQUES FOR VLSI

The high growth rate of circuit complexity of integrated circuits, which is expected to be 40-50 percent/year, influences layout design in two aspects : the increase of computer time due to the slower improvement rate of computer performance (about 25 percent/year) and the decrease of component density. In this section, layout techniques to overcome these two difficulties are discussed.

Layout Model

Layout design of VLSI circuits requires (i) division of a circuit into modules of manageable size, and (ii) treatment of larger modules such as ROMs and RAMs. Therefore, the cellular approach of the previous section should be expanded so as to deal with cells of arbitrary sizes and shapes (general cell). Major problems added to be solved in this layout model are as follows :
1) Placement of nonuniform rectilinear cells into a rectangular chip with minimum area,
2) Routing in channels with irregular shapes, and
3) Single layer routing of power lines.

The placement is the least matured design item. Some of the methods developed so far are the min-cut method (9), the bottom-up construction method of COMPEDA-GAELIC and the AR method in which overlap of cells is eliminated by repulsive force (10). Rather than relying on a single method, the combination of the top-down floor plan step and the bottom-up fine adjustment step of cell shapes and channel width will produce better results (Figure 1). Main components required of a CAD system to support this approach are modular program constitution, database and interactive tools, so as to enable a smooth data flow between various design phases.

In routing, channel area is at first divided into rectangles. The loose router then determines a set of rectangle channels for each net to pass. When the detailed routing is done by the channel router, a proper sequence of rectangle channels should be determined beforehand so as to avoid the occurrence of unroutable channels (11). Another approach is to use the switchbox router at intersections of vertical and horizontal channels (12). A difficulty of this router is that 100 percent interconnection is not guaranteed.

Improvement of Computational Efficiency

The computational complexity of layout design methods is empirically of the order of $O(n)$ at most, where n is the number of components. Considering that it is unlikely that a new algorithm with excessively higher efficiency appears, the effort should be directed to improve hardware performance. An approach is to develop a special-purpose hardware for layout procedures. Most machines reported so far are for routing. The speedup of more than two orders of magnitude can be attained by these machines. This benefits the improvement of layout quality such as component density and circuit performance due to quick iteration cycles. A shortcoming of present machines is the lack of performance flexibility for their costs.

Improvement of Density

The key to increase component density is to decrease area for interconnections. This will be realized most distinctly by increasing metallization layers. The recent progress of process technologies has made three and even four layer metallization feasible. The followings are two examples of the density improvement of gate arrays by the multilayer metallization :
1) The interconnection area for three layer metallization has decreased by 25-35 percent in comparison with two layer metallization (13).
2) The component density by four layer metallization has increased by from 20 to 130 percent over three layer metallization (14).

Routing for three and four layer metallization can be done by the expansion of mothods for two layers.

CONCLUDING REMARKS

Automatic layout techniques of integrated circuit chips have been reviewed. Placement which is the key to successful design still leaves room for further improvement compared to routing. The progress of layout automation can be achieved by the harmonization of three factors, model, algorithm and CAD system configuration.

In the era of VLSIs, turnaround time will be regarded as more important than component density. It should be noted that the large part of layout design is taken up not by the computer time but by the outbreak of redesign cycles. A major cause of design failures is the change of circuit performance after layout due to the addition of undesirable stray elements. This reveals itself after the completion of layout or even after the sample fabrication. Experienced designers take the influence of layout patterns on the electrical performance into consideration even in the course of layout design. Therefore, such techniques are required as to enable intimate interactions between layout information and circuit performance information.

ACKNOWLEDGEMENT

The author wish to thank Mr. H. Sasaki, Dr. H. Kawanishi and the members of CAD department of NEC for their encouragement and useful suggestions.

REFERENCES

1. Hanan, M., Wolff, P.K., and Anguli, B.J., 1973, Proc. 13th Design Automation Conf., 214-224.

2. Matsuda, T., et.al., 1982, Proc. 19th Design Automation Conf., 802-808.

3. Heller, W.R., Mikhail, W.F., and Donath, W.E., 1977, Proc. 14th Design Automation Conf., 32-42.

4. Kato, H., et.al., 1983, Proc. Custom Integrated Circuits Conf., 19-22.

5. Hashimoto, A., and Stevens, J., 1971, Proc. 8th Design Automation Workshop, 155-169.

6. Sato, K., et.al., 1980, Proc. 17th DesignAutomation Conf., 22-31.

7. Horiguchi, S., et.al., 1982, ISSCC Digestof Technical Papers, 54-55.

8. DeMan, H., 1982, "Design Methodologies for VLSI Circuits", Sijthoff & Noordhoff International Publishers, The Netherlands, 173-225.

9. Lauther, U., 1979, Proc. 16th Design Automation Conf., 1-10.

10. Ueda, K., Sugiyama, Y., and Wada, K., 1978, IEEE J. Solid State Circuits, SC-13, 716-721.

11. Kawanishi, H., et.al., 1973, 7th Asilomar Conf. on Circuits, Systems and Computers, 119-124.

12. Chen, N.P., Hsu, C.P., and Kuh, E.S., 1983, VLSI' 83, 37-44.

13. Saigo, T., et.al., 1983, ISSCC Digest of Technical Papers, 156-157.

14. Brenner, S., et.al., 1983, ISSCC Digest of Technical Papers, 152-153.

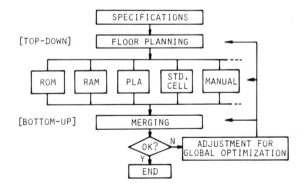

Figure 1 Layout design of VLSI

TABLE 1 - Placement and routing methods for gate array layout.

System	Company	Placement	Routing
—	Fujitsu	Min-cut method and pairwise interchange	Channel routing with new vertical feed-through assignment
PHILO	IBM	Clustering, group allocation and pairwise interchange	Channel routing and clean-up maze
ALEPH	Hitachi	Clustering, group allocation and net balance linear arrangement	Channel routing and maze
GARDS	Silvar-Lisco	Pairwise interchange	Line search extendable to maze
MERLYN-G	VR Information	Force-directed relaxation	Channel routing
MASTER2	NEC	Clustering, group allocation and generalized ralaxation	Channel routing and maze
LAMBDA	NEC	Pairlinking and generalized relaxation	Maze, rip-up and rerouting

TABLE 2 - Computational complexity of typical placement and routing methods.

	Method	Complexity
Placement	Clustering	$O(b^3)$
	Pairwise Interchange	$O(b^2 \cdot r)$
	Min-cut	$O(b^2 \cdot g \cdot r)$
Routing	Maze	$O(n \cdot g^2)$
	Line Search	$O(n \cdot g^2)$
	Channel Routing	$O(n \log n + n \cdot g)$

b: Number of cells
n: Number of nets
r: Number of iterations
g: Number of grids on one side

TABLE 3 - Placement and routing methods for standard cell layout.

System	Company	Placement	Routing
LTX	Bell Labs.	Pairwise interchange	Channel routing (Dogleg router)
CALMOS CAL-MP	Silvar-Lisco	Clustering and force-directed relaxation	Channel routing with cyclic constraint elimination
ALPHA	NTT	Floor plan by AR method, linear and 2D placement, and pairwise interchange	Loose routing and channel routing
ALEPH	Hitachi	Clustering, pairwise interchange and net balance method	Channel routing and maze with optimal feed-through assignment

HARDWARE SPECIFICATION - A USE FOR HARDWARE DESCRIPTION LANGUAGES?

J.D. Morison, N.E. Peeling, T.L. Thorp

Royal Signals and Radar Establishment, Malvern, UK

1 INTRODUCTION

In 'top-down' design a designer often starts with an abstract idea of 'what' a system should do; we call this the 'behaviour' of the system. He refines the design by adding increasingly more information on 'how' to achieve the desired behaviour; we call this the 'implementation' of the system. The sum total of behaviour and implementation makes up the system description. There is a need to describe both behaviour and implementation as unambiguously as possible. We call all such attempts 'specifications'. If the descriptions are truly rigorous, with a mathematical, axiomatic basis we add the qualifier 'formal' to the word specification.

There is not an easy split between behaviour and implementation. For example, a set of boolean equations in 'sum of product' form is a formal specification of a boolean mapping and hence could accurately be called a formal behaviour specification. Such equations can however be mapped directly onto a PLA structure and so contain enough implementation information for someone to make hardware that implements that behaviour, and so they could also be called an implementation. Similarly, the mathematical definition of a finite state machine can be mapped onto hardware consisting of a state register and combinatorial logic. Although there are numerous such grey areas there is no doubt that designers do distinguish between what their designs should do, and how they should do it. To resolve the boolean equations problem the designer must give his reasons for producing the equations. This indicates that the difference between behaviour and implementation can sometimes only be resolved by recourse to the original designer's intentions.

In this paper we start by considering the meaning and uses of behaviour specification. We shall then do the same for implementation specification. The capabilities of Hardware Description Languages (HDLs) for the two different sorts of specification will be considered, and finally we shall examine one particular HDL (Morison et al (1)), with which we are very familiar, to illustrate its strengths and weaknesses as a specification language.

2 BEHAVIOUR SPECIFICATION

Consider the Customer/Contractor relationship. Suppose the customer has a system or subsystem that he wishes the contractor to design and build for him. The customer has a very good idea what the piece of hardware should do but wishes to allow the contractor considerable freedom in how he goes about satisfying this requirement. In this situation the customer knows the behaviour of the system and the contractor has to add the implementation information. There are two uses that the customer might find for a behavioural specification of his requirement.

First, he could use it to transfer his knowledge of the required behaviour to the contractor. This is effectively a person to person transfer of information, so it needs to be both precise and understandable. Natural languages, such as English, plus diagrammatic information, such as timing diagrams, have been used for this purpose but the ill-defined nature of natural languages can easily lead to misunderstandings.

A second use for a behavioural specification would be as part of a contract between the customer and the contractor. Here the customer provides the behavioural specification so that the deliverables of the contract can be checked against it to see if they are behaving correctly. Obviously the behaviour specification needs to be as precise as possible, but the requirement for it to be understandable is much reduced. It is for this second purpose that a formal behaviour specification is particularly attractive because such a specification could theoretically be made unambiguous. The checking of an implementation to see if it satisfies the specification would ideally be done formally. Obviously this requires the specification to be formal, and in practice probably requires the contractor to have done his implementation using tools that encouraged the production of a formally correct solution (rather than doing the implementation using 'traditional' techniques and then trying to do an a posterori proof that it is correct). We do not think that the current state-of-the-art allows such formal checking for anything other than the most trivial circuits, and as a result the checking can only be done by an informal validation process, for example by using extensive test patterns. Even if checking is done informally it does not reduce the requirement to specify the behaviour as accurately as possible, and hence formal specifications are of use even if informal validation methods are used to check implementations against it.

We finish this section with some observations about behaviour specifications. The first two are both concerned with the understandability of the specification.

To make the specification understandable it may well be necessary to structure the information, if only to reduce the amount of detail to be assimilated. For example, a behavioural specification of a microprocessor which describes outputs as a function of both the input stimuli and time, all done at the bit level, is unlikely to contain the sort of information that will allow someone to implement it easily. If however the inputs are described at a less basic level, such as instructions in mnemonic form and data in a more understandable number representation, and if some structure such as the idea of store, registers, stack, multiplier etc is used, then it will be easier for the implementor to grasp what the microprocessor is all about. It would however be naive to suggest that the way a specification is structured will never influence the implementation strategy, for example, the structured behavioural description of the microprocessor might encourage an implementation that uses a hardware stack, store, registers etc.

Our second observation is an expansion of the comment that when a behaviour specification is used as part of a 'contract', the requirement for understandability is much reduced. This is only true in so far as the specification is the template which the implementation must be shown to fit. Noone can prove that the specification is correct. At the very least the specification must be easy enough to understand for the customer to be able to convince himself that he is asking for what he wants.

Our final observation concerns the completeness, or otherwise, of behaviour specifications. It may be important to know whether the specification is complete, i.e. if it specifies the outputs for all possible input stimuli, for all time. It is not however always desirable that such a specification be complete. For example, the specification of a Binary Coded Decimal to Gray code converter might quite sensibly not specify the effect on the numbers equivalent to 10-15. Incompleteness is one thing that distinguishes behaviour specifications from implementations; a corollary is that there can be many implementations that behave differently but satisfy the same specification.

3 IMPLEMENTATION SPECIFICATION

Implementation information for hardware is essentially a physical network of components, where the behaviour of the components is specified either by a behavioural specification language, or as a sub-network of components, or by saying that the component is primitive (in the sense that it is stated to be implementable).

It can be seen that like behaviour specifications implementations can be made more understandable by structuring them. This structuring can both be physical in the sense of grouping together related components as a subnetwork, but also by being able to express the signal values that traverse the network at an abstract as well as a detailed level.

4 HDLs FOR HARDWARE SPECIFICATION

HDLs were originally conceived as validation and design aids, hence they are naturally biased towards describing implementation information. They are however being seriously suggested as possible behaviour specification languages, as can be seen by the requirement for such capabilities in the proposed DoD VHSIC HDL (Preston (2)). The role of HDLs as specification languages is certainly worth investigating because if they can describe behaviour adequately then their restricted and well-defined syntax and semantics would certainly make them attractive by comparison with natural languages. Their origins as description languages for hardware would seem to make them more appropriate than conventional programming languages, such as Ada, Pascal etc., which share the property of a well-defined syntax and semantics.

The current crop of HDLs are not formally derived languages, in that their designs were not based on formal (i.e. mathematical) models of hardware's structure and behaviour (including timing). This does not mean that there is no possibility of defining such models for them. In practice, an 'after the event' formalisation is less likely to yield a model that is easy to manipulate than one that was purpose-made. An attempt to formalise an existing HDL might produce a model that was capable of being analysed (even if not ideally suited to that purpose) but it could also yield useful insights into the problems of designing a formally derived language.

There is a problem in trying to draw general conclusions from the diverse collection of HDLs available. Despite the difficulties it is instructive to detail some of the different solutions to the same underlying problems, even at the risk of over-simplification.

4.1 HDLs for Implementation Specification

We deal with this first because it is the HDLs' most natural domain.

We summarise the HDLs' capabilities as follows:

- Most such languages have comprehensive network construction facilities.

- Most have the capability for describing physical hierarchies, whereby related components can be described as a sub-network which then becomes a component at the next highest level of the hierarchy.

- Some languages allow structure to be given to the signal values that traverse the network. At the lowest level this allows logically related signals to be grouped together. At the highest levels, abstract datatypeing allows signal values to be described at a degree of abstraction far removed from the bit level, where the values can take on a considerable amount of semantic meaning (c.f the comments on structuring the inputs for the behavioural specification of the microprocessor in Section 2).

4.2 HDLs for Behaviour Specification

A behavioural description describes outputs as a function of both inputs and time. Most HDLs deal with time separately. There are three main ways that the current HDLs map inputs onto outputs:

- By the use of inbuilt primitive components whose behaviour is determined by the algorithm in the simulator, for example NAND, NOR etc.

- By including a mapping facility that allows the user to build up his own primitives, to whatever level he chooses.

- By using a separate language for describing behaviour. This can be either a serial programming language (such as Pascal) with extensions that allow it to interface to the timing model, or it can be a Register Transfer Language (RTL) that provides registers to hold internal states. RTLs usually allow transfers to registers as a result of both time and data dependent tests.

There are two main approaches to the description of timing/concurrency:

- Separating out the control structure by allowing the language direct access to the simulator clock. This can control when outputs from pieces of the behavioural programming language are made available to the network, or it can control assignments to registers in an RTL.

- Associating delays with certain parts of the network and then deriving timing properties from the network.

Some HDLs contain both methods of timing control.

Separating out the control structure has the attraction that it divorces timing from the network, in other words the timing is not part of the implementation. The timing is given in a sequential fashion which is fairly natural to experienced programmers. In addition, explicitly giving the events when the state of the machine changes tends to make simulators easy to write (and efficient in their operation).

There are two main problems associated with a separate control structure. The first centres on the complexity that can be encountered when dealing with 'real-life' problems. The timing information can become difficult to unravel and details such as the effect of multiple assignments to the same register at the same time point become very important, and often can only be detected at simulation time (see p.5 and p.18 et seq. of (2)). It is possible that this is a result of the fact that the timing is given sequentially which although it is attractive to programmers is not really the way hardware works.

The second major criticism is that a separate control structure, particularly the RTL model, encourages a particular implementation route - clocked, with heavy use of registers.

Deriving the timing information from the network is much closer to the way the hardware works, and hence is quite natural to hardware designers, although experienced programmers may find it less easy to handle.

Not separating out the control structure links timing behaviour to the particular implementation represented by the network, although it avoids the criticism of particularly favouring one sort of implementation strategy. If such an HDL is to be used for system specification it may become necessary to adopt a set of conventions that certain parts of the network express timing information instead of being part of the implementation. Obviously if the HDL is being used to describe behaviour, with no implementation implications, then the whole network will be there for timing purposes. In this context it might be said that this 'virtual' structure is there to structure the behaviour specification so as to make it more understandable (c.f. the

example of the microprocessor in Section 2).

We finish this section with two comments. First, all current HDLs deal with the values as signals at a single point in time and do not handle time sequences of values (not to be confused with waveform description languages which are solely for driving simulators). Such time sequences would allow the language access to values from previous points in time and so could be used to express internal states. Such a facility might reduce the need for registers in RTLs, and networks that exist only to store internal states.

Our second comment is that in this section we have mentioned two circumstances where a circuit's behaviour can only be determined if the user has some understanding of a simulator's implementation (built-in primitive functions, and multiple register assignments). When a HDL is being used for specifying behaviour it is unfortunate if it is also necessary to describe a particular simulation strategy.

5 ELLA

Having made general comments on the suitability of HDLs for use as specification languages we now examine one such language, ELLA (1), in more detail. There are many aspects of the ELLA system which are vital to its use as a design aid, but which are not considered here in the context of specification languages. These include the associated simulator and the programming support environment with its database/library system, incremental compilation, interfaces to other Design Automation programs etc.

5.1 ELLA Signals

ELLA only handles values at a single point in time. There is considerable freedom in how these values are defined by using the abstract datatypeing facilities available in ELLA (see (1) for full details of the typeing mechanism). The ability to describe signals at varying levels of abstraction allows the designer to start with a very high level view of his system where the signals have semantic meaning (integers, instructions, multi-way control signals etc.) and he can progressively refine these signals until the complete mapping onto bits is defined. We illustrate the use of abstract datatypeing by creating a typeing model for use in a fairly high level description of the AMI S2811 (AMI (3)). We start by defining the basic values that will be made up into the instructions:

```
TYPE
hex = NEW (h0|h1|h2| ......|he|hf|hx|hh),
op1 = NEW (noop|llti| ..... |rept),
op2 = NEW (nop|abs|neg|shr| ......|spu),
base = NEW b/(0..31),
displ = NEW d/(0..7),
shortdispl = NEW (d0|d2|d4|d6),
instaddr= NEW i/(0..255),
```

Our representation of hexadecimal (hex) includes high (hh) and unknown (hx). 'op1' and 'op2' are opcodes. 'base' 'displ' and 'shortdispl' are addresses for the data RAM/ROM. 'instaddr' is the address for the instruction ROM.

From these basic values we can create the following structures:

```
lit   = [3] hex,
fmuvs = (op2,op1,displ,displ),
fmd   = (op2,op1,base,shortdispl),
fmdt  = (op2,op1,instaddr),
fml   = (op2,lit),
```

These create the basic instructions 'fmuvs', 'fmd', 'fmdt' and 'fml' (note that 'lit' is a literal value that is part of the 'fml' instruction).

Finally we describe the general instruction which is one of the four basic instructions, with the added information that one of them (fmuvs) has two variants. We also include an unknown (formx).

```
instformat = NEW (formuv  & fmuvs |
                  formus  & fmuvs |
                  formd   & fmd   |
                  formdt  & fmdt  |
                  forml   & fml   |
                  formx).
```

5.2 ELLA Network Description

ELLA has a hierarchical network description language. The network can be constructed very explicitly using a netlist style of programming. In addition there are features that allow a functional programming style to be used to generate the network. The two styles can be intermixed and a user's initial style will largely depend on his past experience. There is also a facility that allows components to be parameterised, for example it is possible to describe a component consisting of N 4-bit slices. The example below demonstrates the functional way of network construction:

Matrix multiplication of vector 'vec' of length 'n' by a 'mxn' matrix 'mat' gives a vector 'Y' of length 'm' whose qth component is:

$$Y_q = \sum_{p=1}^{n} mat_{qp} \cdot vec_p \qquad \ldots\ldots(1)$$

We shall see that in ELLA this can be expressed equivalently, but a little less compactly as:

```
SIGMA{n}([p=1..n](mat[q][p]*vec[p]))
```

In the first expression $\sum$ • etc. are well known mathematical terminology. In contrast, SIGMA etc. in the ELLA description are not built into the language and have to be defined. In the example below the components of 'vec' and 'mat' are positive integers in the range 0..63 (shortint) and the components of 'Y' are positive integers in the range 0..500000 (vlongint). The symbol * is defined to be integer multiplication of two 'shortint' giving a 'longint' (range 0..3969). The parameterised function (ELLA macro) SIGMA is defined recursively and is specified to be an adder of n 'longints' giving a 'vlongint'. The macro MATMULT delivers a n-vector of 'vlongints' from the multiplication of a n-vector of 'shortints' and an 'mxn' array of 'shortints'.

```
TYPE shortint = NEW shin/(0..63),
     longint  = NEW lngin/(0..3969),
     vlongint = NEW vlngin/(0..500000).

FN * = (shortint:i1 i2) -> longint:
     ARITH i1*i2.

MAC SIGMA{INT n} = ([n]longint:ip)->vlongint:
BEGIN
FN ADD =(longint:i1,vlongint:i2)->vlongint:
        ARITH i1+i2.
OUTPUT IF n=1 THEN ip[1] ADD vlngin/0
        ELSE ip[n] ADD SIGMA{n-1}ip[1..(n-1)]
        FI
END.

MAC MATMULT{INT n m} =
([m][n]shortint:mat,[n]shortint:vec) ->
                                [m]vlongint:
[q=1..m]SIGMA{n}([p=1..n](mat[q][p]*vec[p])).

FINISH
```

Points of detail to notice in the example include:

(a) [q=1..m] is a compact way of writing a repetitive structure of m fields. For example, the body of MATMULT is equivalent to

```
(SIGMA{n}(mat[1][1]*vec[1] +
 mat[1][2]*vec[2] + ... + mat[1][n]*vec[n]),
 .
 .
 SIGMA{n}(mat[m][1]*vec[1] +
 mat[m][2]*vec[2] + ... + mat[m][n]*vec[n])
)
```

(b) ip[1..(n-1)] in SIGMA is a compact way of writing (ip[1],ip[2], ... ,ip[n-1])

The main purpose of this example is to show that such an ELLA program can can be interpreted in two ways. First, it can be interpreted as a parallel programming language creating the vector $\underline{Y}$ of length 'm' whose qth component is Y_q (eqn (1)).

Secondly, it can also be interpreted (figure 1) as a circuit consisting of 'mn' multipliers *, 'm' adders SIGMA (each consisting of 'n-1' adders ADD). The first interpretation could be said to be a specification of behaviour while the second is nearer to an implementation of that behaviour. We suggest that in the end it is only a convention between the writer and the reader which is the 'correct' interpretation.

5.3 ELLA Behavioural Primitives

ELLA does not have built-in primitive components, instead it includes a few behavioural constructs. These constructs can not only be used to create primitive components but they can also be incorporated in high level descriptions. We will demonstrate this by giving some examples of the most important construct, the CASE clause. The CASE clause is a true specification primitive in that it allows a mapping of inputs to outputs with no implied structure. For example the BCD-Gray code converter could be specified by:

```
TYPE b1 = NEW (t|f|x).

FN BCDTOGRAY = ([4]b1:bcd) -> [4]b1:
CASE bcd OF
     (f,f,f,f): (f,f,f,f),
     (f,f,f,t): (f,f,f,t),
     (f,f,t,f): (f,f,t,t),
        .
        .
     (t,f,f,t):(t,t,f,t)
ELSE x
ESAC
```

Note that the ELSE part implies that BCDTOGRAY delivers undefined (x) for the inputs equivalent to the numbers 10-15.

In keeping with other features in the language the CASE clause can be viewed either as a programming language construct or as a hardware construct. As a hardware construct it can be thought of as a very general form of multiplexer. To illustrate the difference consider the interpretation of an ALU's op codes from an instruction. The opcodes are selected by the function PICKOP and are decoded by a CASE clause whose inputs include the opcode (which is the output of PICKOP), the outputs from the accumulator (acc) and from a multiplier (multop):

```
CASE PICKOPinstruction OF
     add: accADDmultop,
     .
     .
     neg: TWOSCOMPacc,
     .
     .
ESAC
```

This CASE clause can be interpreted as a programming language construct which tests the opcode against 'add', ..., 'neg', ..., producing the value after the appropriate colon as the result of the CASE clause (for example, if the opcode has the value 'neg' the CASE clause produces the result of 'TWOSCOMP acc' as its result). The value that the CASE clause produces can be used in another CASE clause, or as the parameter of a function, or wherever else the value would be sensible. This illustrates the use of the CASE clause when viewed as a parallel programming language construct. Figure 2 shows it interpreted as a hardware construct.

5.4 Register Transfer and Events in ELLA

Another example of the use of the CASE clause is one way of expressing register transfer in ELLA. Here 'condition' is taken to be a Boolean value, 'data' is the new value to be loaded into the register 'reg' if 'condition' is true:

```
JOIN CASE condition OF
          true: data
     ELSE reg
     ESAC
-> reg.
```

Here JOIN can be interpreted as LOAD. The input to reg can only be JOINed once so multiple register transfers (assignments) cannot be made (see the discussion in Section 4.2).

If required any desired event mechanism can be constructed within the language. For example, components (operators) AND, WHILE, RISESWITHIN{n} etc. delivering boolean values could be defined so that the following could be written:

```
CASE (aWHILEb) AND (cRISESWITHIN{7}d) OF
     true: ...
     false: ...
ESAC
```

The abstract datatypeing in ELLA allows the difference between 'real' and 'virtual' signals to be expressed quite clearly. For example 'real' digital signals could use the values 'high' and 'low', while the 'virtual' boolean values, used by the event functions, could use the values 'true' and 'false'.

6 CONCLUSIONS

There is a real need to specify both the behaviour and implementation of hardware. Formally derived languages offer the best hope for the future; such languages are however still a research topic and it is unlikely that any such language will be commercially useable in the next few years. In the meantime, HDLs are sufficiently advanced to be useful for hardware specification. They are well suited to specifying implementations and the best of them are preferable to natural languages and conventional programming languages (e.g. Ada) for specifying behaviour. It is likely that using HDLs for specifying behaviour will require the development of new techniques in their use to suit them for their new role.

Efforts to formalise existing HDLs should be pursued and will serve two purposes. First, it will make them more suitable for specification, and secondly, it could provide useful experience that will aid in the development of purpose-built formal specification languages.

7 ACKNOWLEDGEMENTS

The authors would like to thank Ian Currie, Michael Foster and Stephen Goodenough for their help in writing this paper.

8 REFERENCES

(1) Morison J.D., Peeling N.E., Thorp T.L., 1982, ELLA: A Hardware Description Language, IEEE Int. Conf. Circuits and Computers, New York, pp 604-607.

(2) Preston G.W., 1981, Report of IDA summer study on HDL, Oct. 1981, AD A110 866.

(3) AMI, 1981, MOS Product Catalogue.

Crown Copyright © 1983

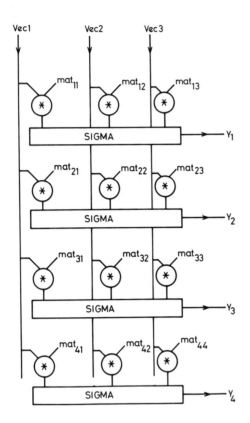

Figure 1   Circuit given by MATMULT

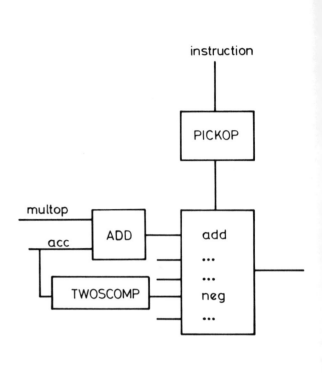

Figure 2   CASE clause - opcode decode circuit

VLSI TESTING - ITS POTENTIAL AND PROBLEMS

T W Williams

IBM Corporation, USA

ABSTRACT

With the vast increase in VLSI designs the area of testing provides an excellent environment to help manufacturers gain a competitive edge. The reason for this is the great demands on quality and reliability that is being required in the marketplace. The potential for the future will come with some built-in problems. One problem, clearly, is complexity which has a history of doubling every two years. The second problem is the flexible environment which is required to attain high quality levels. A key example of this rapidly moving industry is the fact that yesterday gate array vendors and silicon foundries were unheard of. Today, they are an integral part of the industry and growing. One can even find brokers of second level packages. Without flexibility in all areas, this industry would not have experienced the growth it has seen.

This presentation will give a detailed view of the techniques used in the testing industry, as well as the techniques which will be employed in the future. It will show how research activities flow into the production areas, retaining their flexibility in order to secure a place in future technological evolution.

TESTABILITY ANALYSIS

William L. Keiner

Naval Surface Weapons Center
Dahlgren, Virginia   USA

INTRODUCTION

As VLSI circuits become more complex and more critical to system operation, there is an acknowledgement that new designs must be somehow constrained to be more easily tested or they may become completely unsupportable. As a result, Design for Testability is becoming increasingly important to the Services and industry.

One of the keys to a successful testability program is the ability to assess the testability of a proposed VLSI design early enough in the design process to allow redesign, if needed, at minimal cost. Several automated tools have been developed to assist in testability analysis.

This paper describes some important testability techniques for VLSI design and discusses the various kinds of testability measures being used. The paper concludes with an example of a Navy Testability Analysis program and a description of the new Navy standard Automatic Test Generation program.

VLSI TESTABILITY APPROACHES

A great variety of testability approaches may be found in current VLSI designs, both in the military and commercial sectors. However, almost all can be categorized into a small number of generic approaches.

Most important are those techniques which electrically partition, in test mode, a large, complex VLSI structure into a number of smaller, easier-to-test substructures. The critical issue here is the ability to create relatively simple interfaces between partitions which may be controlled and observed by the test process.

Set-scan appears in VLSI designs under a multitude of names. Here, an external tester (or perhaps an on-chip pattern generator) is able to directly set all or most internal latches in a chip through a serial shift register structure. The test patterns, once loaded, are allowed to propagate through the combinational logic, and are clocked into the receiving latches. The same serial shift register structure is used to send the captured test information back to the tester. Set-scan then offers perfect controllability and observability of the VLSI chip and simplifies the test generation process since all that is left to test is the combinational logic between latches. It should be noted that the set-scan approach is limited to static testing situations.

As VLSI packages grow in terms of the number of pins available, there is a recognition that some pins must be dedicated to the testing function. Thus VLSI chips may have six or more pins used for a dedicated test ports to handle test control signals, serial input/output, error signals, etc.

Often the increased capability of VLSI circuits is achieved through large amounts of on-chip ROM and firmware. This provides an excellent opportunity to utilize diagnostic test patterns (microdiagnostics) in available ROM words.

In some situations, effective test patterns may be generated on-chip without sophisticated test algorithms by using pseudo-random pattern generators. This may be achieved using a linear feedback shift register (LFSR) as the source of patterns. A LFSR may also be used to compress the test response data into a "signature" without significant loss of information. The two approaches may be used together to test combinational logic at full clock speeds.

The previous approaches can be used to test the VLSI in the manufacturing test environment or in the "off-line" maintenance environment. Sometimes, it is important to detect errors in an operational environment as they occur or soon after they occur. This kind of built-in monitoring circuitry is usually incorporated in the form of parity and duplicate elements/comparators. The latter structures, though expensive in terms of chip area, can be also used to support fault tolerance, if needed.

TESTABILITY MEASURES

There are three major measures which may be used as predictions of test effectiveness: fault detection, fault isolation and test time. In dealing with VLSI design, fault detection is undoubtedly the most important testability measure, both in a manufacturing environment and in a maintenance environment. Fault detection is usually graded using a fault simulation program and is discussed in the final section of this paper.

As VLSI-based systems are designed to be more testable there is a natural tendency to use the chip interconnect structure (e.g., busses, test ports) to facilitate fault isolation. Thus it is important during system test to be able to (1) select each chip, in turn, and (2) apply comprehensive fault detection tests to each chip.

Unfortunately, the answer to the question "Is the fault detection comprehensive enough?" usually is not known until relatively late in the VLSI design cycle, after test software development. If the test is not comprehensive enough it is much too late to impact redesign of the VLSI circuit for improved testability. It is for this reason that a concept called "inherent testability" has become important.

Inherent testability may be defined as a unit's potential to support comprehensive, fast and low-cost testing. It is based upon the design of the unit and is independent of test pattern design.

Inherent testability measures may be categorized as three types: (1) rule checkers, (2) ad hoc checklists and (3) algorithmic approaches. The type of measure used depends primarily upon the evaluation environment.

If a highly structured design for testability technique (e.g., the IBM Level-Sensitive Scan Design) is imposed on a VLSI design, the best measure may simply be an audit of how well the pre-defined design rules are followed. Rule checking is an excellent method for situations in which DFT is totally controlled, such as a contractor's evaluation of his own design.

A totally different situation exists if the testability evaluation is being performed by a Defense agency. Here, the agency has little or no control over the contractor's detailed design and must rely upon more general evaluation approaches. A proposed military standard entitled "Testability Program for Electronic Systems and Equipments" was developed by the Navy and provides a core list of testability attributes. The standard requires that the contractor add, delete or modify items in the list and assign weights to each. Here, the contractor and the government are forced to agree relatively early in the development as to exactly what the testability priorities are to be. As the design progresses, the contractor is scored on how well the agreed testability is being addressed. Structured testability approaches may be easily included in this approach. The approach may also be utilized by prime system contractors in dealing with subsystem and device subcontractors.

Although the weighted checklist yields quantitative data, the approach is essentially qualitative. In recent years, a number of computer programs have been developed to provide a more objective, algorithmic approach to the measurement of testability. A sampling of these programs is listed below.

   Testability Measure (TMEAS)

   Computer-Aided Measure for Logic
     Testability (CAMELOT)

   Sandia Controllability/Observability
     Analysis Program (SCOAP)

   Inherent Testability Figure of Merit
     (ITFOM)

   Controllability-Observability-
     Predictability-Testability
     Report (COPTR)

   System Testability and Maintenance
     Program (STAMP)

These programs were written for the early testability analysis of digital systems and most have been applied to VLSI design. As an example of the above, the Inherent Testability Figure of Merit (ITFOM) approach is discussed in the next section. ITFOM was developed by Sperry Univac for the Navy in 1981 and is based upon the Sperry TESTSCREEN algorithms.

INHERENT TESTABILITY FIGURE OF MERIT (ITFOM)

In ITFOM, a multidimensional testability measure is derived using controllability and observability calculations for each logic node. A node is either a simple gate or a single output functional element. A functional element with N outputs results in N nodes. Controllability of a logic node is the ability to control (set) that node to logic high and low by choice of circuit input values. Observability of a logic node is the ability to propagate the state of the signal at that node to an output pin where it can be externally observed.

The process of test pattern generation is a sequence of controlling and observing tasks. For instance, in order to detect a stuck-at failure on a logic line, such a line must be controlled (set) to the opposite value of the faulty condition; next, a sensitizing path must be established toward an observable output pin. Both of these activities require many controlling tasks for other contributing logic nodes in the circuit. If these tasks are easily achievable, then the testing process is likely to be easily accomplished. The problem is one of how to quantify this controllability and observability.

A controlling or an observing task requires certain primary input pins to be constrained to logic high or low values. The number of constraints on primary input pins gives a measure of conflict probability. Controllability high or low of a node output is defined as the number of network primary input pins which must be constrained (set high or low) in order to control that node output to logic high or low, respectively. Similarity, observability of a node signal is defined as the number of network primary input pins which must be constrained in order to observe a fault on that node at a primary output. The observability value of a fanout line is equal to a minimum of the observability values of the nodes to which it fans out. Note that a smaller observability value implies better observability. The observability of a node input is determined as a function of the observability of the node output and controllability of other node inputs.

The concept of time frames is necessary to handle sequential elements. Associated with each combinational value is a corresponding sequential value indicating the number of clock changes (time frames) needed to control or observe that logic signal.

The ITFOM procedure for calculating controllability/observability values of a logic circuit involves using a library of routines; one for each gate and functional element type. Starting at the primary inputs of the circuit, controllability values are calculated at the output of each node, based on controllability of the node inputs. Controllabilities are calculated progressively through the circuit until all nodes have been evaluated and all primary output pins have assigned values. Observabilities are then calculated starting at the circuit outputs and working backwards, since node input observabilities depend on output observability and input controllabilities.

ITFOM carries out these controllability/observability calculations and prints out the values of each node. There are six columns associated with each node name. Columns 1 and 4, 2 and 5, and 3 and 6 give dependent linked pairs representing controllability low, controllability high and observability, respectively. The first column in each pair represents combinational values and the corresponding higher column gives the sequential counterpart. Lower values imply easier controllability or observability. The sequential values given indicate the number of clock changes necessary to control (observe) the signal, which implies the number of test patterns needed. Combinational values give a measure of probability of conflicts.

AUTOMATIC TEST GENERATION (ATG) AND SIMULATION

The Navy has recognized that their current digital simulation programs would not be adequate for generating tests for VLSI/VHSIC complexities and grading fault coverage. Thus a new Navy standard ATG called HITS (Hierarchical Integrated Test Simulation) is being developed by Grumman Aerospace Corporation to meet anticipated needs.

As digital circuits increase in size and complexity, new techniques have evolved to keep ATG memory requirements and system processing speed within acceptable limits. This has been accomplished by introducing higher level functions in place of their structural equivalent circuits. This approach has resulted in a drastic reduction in node count and memory requirements, and therefore processing speed has increased. While these represent the positive effects of functional models, there is an inherent disadvantage in that internal fault visibility is limited.

Hierarchical simulation approaches the problem of component model definition and internal fault coverage by allowing model simulation at different levels of detail. For example, one extreme for a given component might consist of the Register Transfer Language (RTL) definition of the model, which has the advantage of modeling an integrated circuit without defining the internal structure. However, it provides minimum internal failure data. Another model of the same component may consist of an equivalent gate level structure which includes both the internal structure and fault data. The capability to process multiple versions of a given component is accomplished through the HITS architecture which enables a "model swapping" scheme to be implemented. Hierarchical simulation, therefore, is the technique of starting with a component at the behavioral level (e.g., RTL description) and swapping into the topology data base the more detailed model which incorporates and defines the structure. From this, internal fault information can be derived.

Each time a model is swapped, a new set of failures pertinent to the structural representations will be defined. The detection of these failures will then constitute the objective of the subsequent phases of test generation and test simulation.

When a satisfactory detection figure has been achieved, after having iterated through the test generation phase and the test simulation phase, the user can invoke the swapping of the structural representations of a different set of high-level functions. Then the user can continue on to the test generation and test simulation phases.

The output from the test generation and simulation phases are essentially tests and fault data as they relate to the multi-level characteristics determined by invoking hierarchical simulation.

CONCLUSION

Design for testability has become a crucial part of the VLSI design process for supporting both factory test and maintenance test. Testability analysis tools are being developed to support both early VLSI design analysis (e.g., the ITFOM program) and later test effectiveness analysis (e.g., the HITS program). These tools are by no means in their final form but can make a real contribution toward quantifying progress toward testable designs.

CHAIRMAN'S REPORT - SESSION 3.1 "EXPERT SYSTEMS AND VLSI"

G. Musgrave

Brunel University, UK

INTRODUCTION

This session was conducted in a true 'workshop' fashion where no notes were taken and all those taking part were assured that they could speak freely. Consequentially this report can only mention some of the basic aspects discussed. It must be emphasised however, that there was a great deal of interest in the topic, which continued as a topic for discussion outside this session.

Many people who attended the session where interested to ascertain what are "expert systems". A difficult question to answer in detail but a simple definition can help namely:

An expert system is a program embodying knowledge from an expert skill so that it can offer intelligent advice or take an intelligent (processing) decision and most desirably can on demand, justify its own line of reasoning intelligibly to the end user.

During the whole of the three day seminar, many speakers discussed particular aspects of VLSI, but the underlying problem was always complexity. When decisions cannot be made on facts (they are not available) or algorithms (simple rules cannot apply) then knowledge based systems will gain credence. It is only correct that there are sceptics about this area. There is a significant group of microelectronics workers' who do not believe that the top down, hierarchical approach to design, can produce the working VLSI design and that only simplistic rules at the basic cell level can guarantee a working chip, and therefore experts systems have no place. There are other groups who argue that there can be no experts in VLSI and therefore no foundation for the IKBS system. However, the majority of people were delighted to learn what was going on and it fell to Stephen Rhodes of Racal Expert Systems to provide a basic tutorial.

The use of an expert system does not have to be confined to a computer automated solution on the contrary an expert system can be used to give problem solving assistance to inexperienced professionals as well as being used to refine knowledge for expert users. The examples given by Rhodes were from the oil exploration field where many experts have their own knowledge base and there is a need to consider all the aspects and the many multifunction interactionships to achieve correct results. This can be related across the multi-disciplines of chip design to possibly achieve more effective results than simply having a common database. Another aspect which was usefully debated, was that expert systems can be effectively utilised in a training role. For many members this was a new aspect which held promise for the acute shortage of skilled personnel in this field. This lead to a broader discussion of how expert systems could enhance human capabilities; improve access to knowledge, sharing of knowledge, combining knowledge, mechanism for structuring and refining knowledge and perhaps above all, enhancing the ability to learn.

The workshop then turned its attention to why had these systems started to emerge and what sort of environment was needed. Stephen Rhodes' two slides, on enabling technology, provoked considerable discussion. [They are given here as Figure 1 and 2] But these points of discussion became academic when the debate moved to, should or should it not be, a LISP machine. Clearly the virtues of this symbolic language were extolled.

-side effect free programming (incremental compilation and parallel processing).

-code and data indistinguishable thus program generation.

-proven high programmer productivity for large packages.

However, when the realities of the size of the machine were appreciated and subsequent costs, a number of European delegates become very worried. Very few people took comfort from the fact that once one of these machines was available to a LISP programmer he became super efficient and produced enormous programs very quickly. But then it is the hardware designs who are making this development possible with the higher definition graphics display and con-current processing.

Then a number of speakers outlined what their current work is in this field. Most of the initiatives were coming from the larger organisations and were in their infancy and no-one reported any work to compare with that of Xerox Park and Carnegie Mellon University in this important field of VLSI.

ACKNOWLEDGEMENTS

As chairman of the session and the author of this note, I would like to thank all those who took part and helped to share their expertise without a program.

BIBLIOGRAPHY

The following references may be helpful to those seeking a background in this field.

1. Conway, L. 1981, The MPC Adventures: experiences with the generation of VLSI design and implementation methodologies. Proceedings of the Second Caltech Conference on very large scale integration, 5-28. (Also reprinted as Tech. Rep. VLSI-81-2, Xerox Palo Alto Research Center).

2. Davis, R. 1982 Teiresias: applications of meta-level reasoning. In R. Davis & D.B. Lenat (Eds). Knowledge-Based Systems in Artificial Intelligence. New York: McGray-Hill

3. Duda, R.O., & Gaschnig, J.G. 1981, Knowledge-based expert systems come of age. Byte 6 (9): 238-281.

4. Feigenbaum, E.A. 1977, The art of artificial intelligence: I Themes and case studies in knowledge engineering. IJCAI 5, 1014-1029.

5. Guibas, L.J., & Liang, F.M. 1982, Systolic stacks, queues and counters, Proceedings of the Conference on Advanced Research in VLSI, 155-164.

6. Hayes, P.J. 1979, The Naive Physics Manifesto. In D. Michie (Ed). Expert Systems in the Micro-Electronics Age. Edinburgh: Edinburgh University Press.

7. Lyon, R.F. 1981, Simplified Design Rules for VLSI layouts. Lambda: The magazine of VLSI Design, First Quarter, 54-59.

8. Minsky, M. 1961, Steps toward artificial intelligence. In E.A. Feigenbaum & J. Feldman (Eds). Computers and Thought. New York: Mc-Graw-Hill.

9. Nilsson, N.J. 1982, Artificial I Intelligence: Engineering, Science, or Slogan, The AI Magazine, 3, (1), 2-9.

10. Sahal, D. 1981, Patterns of Technological Innovation. Reading, Mass.: Addison Wesley.

11. Simon, H.A., 1981, The Sciences of the Artificial (2nd Ed). Cambridge, Mass.: The Mit. Press.

12. Stefik, M., Aikins, J., Balzer, R. Benoit, J., Birnbaum, L., Hayes-Roth, F., & Sacerdoti, E. 1982, The organisation of expert systems: A Prescriptive Tutorial. Artificial Intelligence, 18; 135-173.

13. Stefin, M., Bobrow, D., Bell, A., Brown, H., Conway, L & Tong, C. 1982, The partitioning of Concerns in Digital System Design, Proceedings of the Conference on Advanced Research in VLSI, 43-52.

14. Tong, C. 1982, A Framework for Design, Memo KB-VLSI-82-16 (working Paper), Knowledge-based VLSI Design Group, Xerox PARC.

15. Stefik, M., Conway, L., 1982, Towards the Principled Engineering of Knowledge, Xerox Pala Alto Research Centre, the AL Magazine, pp.4-16.

Figure 1 Enabling technology - *software*

- Knowledge Representation
    - Symbol manipulation languages
    - High level descriptions
    - Inference techniques
    - Knowledge-based search techniques

- Man/Machine Interface
    - Explanations
    - Window systems
    - User models
    - Mixed initiative

Figure 2 Enabling technology - *hardware*

- Advanced Architectures
    - Prefetch
    - Cache memory

- Large High Speed Memories
    - Dynamically writable control stores

- Man/Machine Interface
    - High resolution graphics
    - Cursor control devices
    - Windows

CHAIRMAN'S REPORT - SESSION 3.2 "HIGH SPEED VLSI"

F. G. Marshall

Plessey Electronic Systems, UK

The topic of high speed performance from small geometry devices in both silicon and non-silicon technologies produced a lively workshop discussion. The major points to emerge included:

1. The enthusiasm for any technology that would not run at room temperature was minimal. Even IBM have reduced their commitment to Josephson Junction research.

2. Device geometries will continue to reduce. However, there is a growing consensus that there will be little to gain from reducing minimum dimensions below 0.5 µm, at least for standard two-dimensional micro-circuits. This limit is caused by the onset of electron velocity saturation, the problems of driving connector capacitance and the increasing similar effective values of gate capacitance and depletion capacitance. It is fortunate that 0.5 µm devices can be constructed without recourse to electron beam direct slice writing.

3. There was general agreement that for true VLSI, silicon will be preferred to GaAs for several years. Advances in silicon technique, still exploiting the silicon oxide insulator advantage, may well remain a step ahead of GaAs technique.

4. Silicon on insulator would reduce lead capacitance and vertial field problems. A truly three dimensional geometry would offer huge advantages but seemed difficult technically.

5. To reduce the time taken for a calculation, it is not essential to continually increase switching times - parallel processing can also solve the problem. There was a consensus, however, that the demands of moving picture signal processing (for example) were so great that the extremes of both parallel processing and device speed were required simultaneously.

CHAIRMEN'S REPORT - SESSION 4 "CHIP DESIGN METHODOLOGY AND TECHNOLOGY"

P.B. Denyer* and J.B. Clary**

*University of Edinburgh, Edinburgh, Scotland,   **RTI, North Carolina, U.S.A.

This session was really two mini-sessions, each with a strong theme that was well developed by the speakers and actively debated by the audience.

In paper 4.1 John Gray gave a survey of silicon compilation since Rem first coined the term for Johannsen's work at Caltech in 1978. He reviewed three styles of silicon compiler and showed a common theme in structure description languages. John Fox gave a complementary presentation, examining currently successful design methods and stressing design for performance at lower levels in the hierarchy. The audience debate revealed a wide appreciation of the concomitant needs of rapid, high-level design on the one hand and performance per unit area of silicon on the other. Issues such as structural versus behavioural specification, and textual versus graphic input were contested. Interestingly, the notion of highly automated chip design was not challenged.

In paper 4.3 Gilbert Declerck reviewed the limits to performance in low geometry MOS devices, demonstrating a clear tail-off in performance below channel lengths of 0.5-0.8 micron. Gilbert emphasised interconnection as a dominant performance limiter for low geometries. David Grundy in paper 4.4 compared the MOS technologies with bipolar for VLSI applications. He stressed advantages for bipolar in speed and power, though offset and drift marred analogue performance. His conclusion was a broad equivalence between bipolar and CMOS for VLSI. The audience were quick to challenge this view, which David defended admirably against the CMOS 'bandwagon'. There was some discussion of SOS and SOI developments, which several participants advocated as a winning VLSI technology for the future.

Participants seemed to enjoy these sessions for the key issues they raised over new approaches needed for the future.

SILICON COMPILERS

J P Gray

Lattice Logic, Edinburgh, UK

ABSTRACT

Silicon compilation is a subject fast gaining maturity; since Rem first coined the term to describe the work of Johannsen at California Institute of Technology in 1979, at least three new-start companies have been formed which purport to offer software products or design services using this type of tool. This is interesting as, apart from being a notable violation of the normal five-year period between research and production, it is a subject area with no precise definition, no "theory" and only limited practice.

At its simplest, silicon compilation involves describing chip design by writing a computer program in contrast to the normal graphical construction of artwork. Normally, by executing this program description, artwork is generated in a particular design style for a particular process technology. If these program descriptions can be written in an appropriate manner then there are a number of advantages to this approach: a better control of the correctness of a design, a more rigorous optimisation of a design and an approach to porting designs between processes.

The obvious analogy is between high level language compilers translating programs to machine code and silicon compilers translating chip descriptions to layouts. This paper surveys the subject's recent history and explores the software analogy in more detail.

REVIEW OF ARCHITECTURES AND THEIR INTERACTION WITH SYSTEM PERFORMANCE AND DESIGN METHODOLOGY

J. Fox and S. Jamieson

Plessey Research (Caswell) Plc, U.K.

INTRODUCTION

The available chip complexity now offered by VLSI (sub-3 micron) technologies constrains IC designers to consider in detail (silicon) system architectures. VLSI chip design is based on sequential, but inter-related decisions on specification, architecture, partitioning, performance, logic design methodology and technology.

Architectures are highly technology dependent. With VLSI technologies below 1.25 micron feature size, the maximum speed may be limited by interconnect rather than by gates (1,2). Does this alter present design concepts by constraining all designs to minimise communication paths? Present projected interconnect limitations of course assume that architectural and algorithmic research, and metallisation technology do not advance.

To maximise the performance of a given technology then pipelining or parallelism (concurrency) at the bit level should be employed. A penalty in data latching is observed, but the approach minimises the data communication area and speed requirements. General purpose microprocessors cannot take advantage of the bit-level approach. Dedicated single algorithm processors are much more likely to be implemented in such a fashion. However, it is more common to see overall performance determined by functional blocks (RAM, ROM, PLA, multipliers, ALU, etc.) which may have delays of 10-50 times greater than the intrinsic (2-input NOR) gate delay of the process.

We present in this paper a survey of some of the architectural concepts and related design methodologies for VLSI, for both general purpose microprocessors and dedicated DSPs.

VLSI Technology Choice

VLSI technology choice is particularly important as many high speed applications are military driven. In these cases emphasis is placed on radiation hardness. It is well known that bipolar technologies, I2L/ECL are harder than bulk, or even possibly on-insulator CMOS. Although the power-delay product of say I2L is less than the corresponding CMOS process, it is often seen that I2L designs have a significantly higher power dissipation for the same overall functional complexity. This is essentially due to the constant power consumption nature of bipolar structures, as opposed to the dynamic dissipation of CMOS. Of course, clever circuit design techniques such as stacking levels of I2L gates may be employed, but may be application specific. This approach maximises current utilisation, but is not attractive for parameterised, or automatically synthesised cells. High complexity ECL designs are impractical due to power limitations, but do offer sheer speed. However ECL structures may be used in I2L designs to enable bus driving but again these need to be customised; a programmable power option is possible. Note however, that the radiation requirements in such areas as telecommunications are much less severe, but usually such applications exhibit tight power specifications.

Several recent chips have demonstrated clever circuit design techniques and processes, (19,23); note double layer metallisation is now regarded as standard. Extra layers of metallisation will ease the autorouting problem and reduce capacitance to substrate, but at the expense of increased inter-track capacitance due to fringing. Future technological and design methodology advances must be driven by the systems requirements. However, VLSI scaled technologies soon to be available will call for revisions to many current systems standards.

VLSI Chip Design Methodologies

System performance is ultimately technology limited, but it may also be limited by the design methodology adopted and the system architecture. Design considerations often require that the number of functional units is low, their behaviour is fairly simple, and that they are regular and interface well, both functionally and physically. The system architecture may limit the performance by technology effects if, for instance, communication paths are long or convoluted. To maximise performance from the technology a bottom up design methodology is best, ie. from the transistor level upwards. However, for complexity management in large systems it is best to design hierarchicaly, ie. top down.

If system architectures require functional block performance within the capabilities of the chosen technology, then semi-custom methodologies (gate-array, standard-cell) may be successfully applied. Design automation (DA) tools such as auto-place/routing are available in most gate-array systems. In standard-cell, DA tools such as PLA generators (3), ROM and RAM (4) assembly software, dense gate matrix synthesis software are becoming increasingly available. A synchronous design methodology encourages further the use of such DA tools, although the power problems of a dynamic synchronous technology should not be underestimated. These tools and methodologies have been driven mainly by N/CMOS technologies; they are not technology independent.

Pre-design knowledge of production quantities, estimated gate complexity, design time, may have an overriding influence on the design methodology chosen. Gate-arrays are not so well suited to designs with a high degreee of regularity and will continue to be used for 'random' logic. A large requirement of VLSI will be to reduce system package count, ie. technology insertion; gate-arrays may well fill this application. Standard-cell approaches are being increasingly adopted, even for moderate volumes, if the design is structured in any way. Variable size gate-arrays may of course be adopted as standard-cells, or vice-versa (5) to take advantage of available DA tools and to incorporate small amounts of memory.

But what of custom design? Parameterisable (within limits) standard cells may be an answer, where the size and possibly performance of a cell may be altered to fit its situation. System performance is often found to depend only on a small core of the total system. If this core can be optimised, the peripherals may be designed with less attention to speed and power, reducing overall design time. VLSI design criteria are not the same as at the board level. At board level, it is important to minimise the number of components, whereas in VLSI communication paths are equally important. If power dissipation allows, more devices can often be used to reduce the length or complexity of signal paths - as in PLAs. The tradeoff between hardware and software is also different in VLSI systems, and is a balance between managing complexity and achieving the required performance. If standard-cell performance is marginal, then new, optimised cells must be produced (customised). At present, cells are relatively simple, but as cells become more complex then custom design (of cells) will become significant.

VLSI Microprocessor Architectures - An Overview

Processors are a general purpose component, used to support the implementation of systems in software. The use of software to implement systems has become widespread, as it allows any system architecture to be defined and any operations to be performed on this system architecture.

Modern general purpose processors are trying to aid system performance in two ways. The first is to increase the basic processor throughput. The second is to integrate parts of commonly used system architectures on chip. For example, most high end microprocessors for general purpose computer type applications need to run a fairly complex operating system. This operating system architecture can be mapped directly onto silicon, which should lead to an increase in OS performance - and hence system performance. This is the approach taken by many modern processor manufacturers to some degree on high end MPUs. The most extreme example is the Intel iAPX 432 (6,7,8). This is a three chip set, where much of the OS kernel is integrated on chip. It is designed to run the high level language Ada efficiently. The decision to implement the OS kernel in hardware is due to the throughput overhead software mechanisms for the Ada language might have.

Complexity considerations in VLSI systems are leading to wider application of structured design methodologies. This gives similar advantages to high level languages, ie clarity, correctness and ease of modification. It also leads to more clearly defined communication paths, which gives regularity and hence more compact layout, but might give less low level performance than a bottom-up, optimised design. However, higher level approaches lead to more concentration on the algorithms as the limit to system performance. The equivalent in modern processor designs is the use of better system architectures to allow more concurrency, and closer matching of the processor to memory. Concurrency in processors is seen mainly as instruction pipelining. Pipelining in its basic forms (such as overlapping instruction fetch and execute cycles) has been used since the early days of microprocessors. More modern designs are including more sophisticated prefetch schemes, as well as overlapping various stages of the complete instruction execution within the machine. Pipelined machines are generally partitioned into several semi-autonomous sub-machines, each performing a different phase of instruction execution, such as operand/address derivation, translation/protection etc. Examples of this type of architecture are the Intel 80286 (9) (3 levels of pipeline), which is claimed to give up to 6 times the throughput of the 8086, the DEC T-11 (10) (3 independent sub-machines), and to a lesser extent the Bellmac 32/32A (11) and Fairchild F9450 (12). Zilog have also announced that their Z80,000 32 bit CPU will use 6 levels of pipelining to improve throughput (13).

All processors show a disparity between the basic switching speed of the gates available in the process (typically about 2nsec under light loading) and the clock cycle time of the processor (best about 50nsec). This disparity on the electrical performance of processors can be traced to three major causes. First, there is the control cycle of the machine. In a microprogrammed processor, this can be an access time delay on a large ROM or ripple through time on several small PLAs. Second, there is the delay due to driving long buses in the processor datapath, and finally there is the delay due to the operation of the datapath units. The longest datapath unit delay is often the ALU carry propogation time. In general purpose processors, where all registers need to communicate with each other and the datapath units, there is little alternative to buses. It may be possible to speed up datapath units using techniques such as carry lookahead, and to reduce control cycle time by using smaller control stores, fast PLAs or hardwired control logic, but due to the general purpose nature of the processors, bus delay will remain a problem. Note however, that in specialised processors such as signal processing devices (Eg TMS 320 (14)) the communication paths may be relatively fixed, so general interconnection buses may not be necessary. The limiting delay in a programmable signal processor is more likely to be from the datapath units, ie the array multiplier.

The chip architectures of all modern microprocessors show a clear separation between datapath(s) and control section(s). Datapaths are almost always based on the approach described by Mead and Conway for the OM 2 Datapath chip (15). This is the bit sliced technique, with integral buses and butting datapath units. A good example of such a structured datapath design can be found in the TMS 7000, a recent 8 bit microcomputer (16). The control structures of modern machines are more diverse, and show different approaches to achieving high throughput with a regular control structure.

Microprogramming is extensively used to ease the design problem, and does not seem to lead to any less performance than a hardwired approach. The Z8000 is one of the few processors with such a hardwired control section, but its performance is not dramatically different from other microprogrammed machines such as the 68000 or 8086. The Z8000 control section may use less area than an equivalent microprogrammed controller, but must be far more difficult to layout and modify.

Distributed control is often implemented using local PLAs, closely coupled with datapath units, driven by a 'vertical' microcode store. This is the type of approach favoured by Intel in the iAPX 432 series, and also to a lesser extent in the 8086/80286 devices. Alternatively, a wide, horizontal microword can be used, with more direct control over the datapath. This type of controller architecture can be seen in the TMS 99000 16 bit processor (17), the Bellmac 32A and HP FOCUS 32 bit CPU (18). A wide microword has the advantage of less decoding between the controller store and datapath units, but may also give a slower microcode store, and more redundant bits in that store. The horizontal microword has the advantage of simplicity, and the microstore may be minimised.

The difference in control section designs illustrates a difference in design philosophy between the management of complexity, and the reduction of complexity. The iAPX 432 is a prime example of managing complexity, while the designers of the HP FOCUS stated reduction of complexity as a design goal. The design was partitioned into a fairly small number of functional blocks which were custom designed. It is heavily microprogrammed, with a fairly straightforward control and datapath structure. Much of the performance (ie, speed and area) of the FOCUS CPU is due to a clever process, and although the chip also dissipates a large amount of power, this has been judged a good system tradeoff.

Intel's methodology of managing complexity pays off when applied to its newer 80186/80286 highly integrated processors. These effectively include the memory management system and other peripherals such as timers and DMA controllers onto a single chip. This is very attractive for system designers who have already developed systems using the 8088/8086 devices as it retains software capability and reduces device count. The performance of these devices is also increased due to extensive pipelining.

The RISC architects have taken a different approach to improving the performance of a programmable architecture (20). They have kept their instruction set simple and small, and effectively moved many of the general registers onto the chip for higher speed. With a suitable operating system, such a device could possibly be faster than the more complex 432 series, and equally as protected and secure. The fact that an OS kernel is implemented as hardware on the 432 instead of external firmware does not make it any less likely to crash, and indeed it is far more difficult to modify if subtle bugs are discovered - it also makes the chip a lot less flexible.

The RISC is more likely to be a general purpose semiconductor component, suitable for the support of systems in software whereas the iAPX 432 is a good example of a complex system implemented in silicon. It has yet to be seen which approach will prove more practical for this particular application area (ie Ada systems). Consider also that the RISC makes much better use of the technology as it is regular, simple, easy to use and understand. It could be argued that it may be more difficult to program effectively, and more prone to fatal software error, but with modular component software available in firmware this should reduce the programming overhead. The basic performance level (ie. clock rate) is also likely to be higher, as the critical paths are more easy to identify, and as design time is inherently lower, more optimisation is feasible.

Identification of critical limiting paths on the total system performance often leads to a better processor architecture. Two interesting examples where novel techniques have been devised to meet specific system performance requirements are the Hitachi HD6301 (20) and Motorola MC6804P2 (21). The Hitachi device is a CMOS microcomputer designed for very low power applications, so its peripherals (timer, serial interface) operate at a slower clock rate than the kernel CPU, without compromising performance. The Motorola chip is also a microcomputer, intended for very low cost applications. To reduce die size to a minimum, its designers have gone for a serial internal datapath, and increased the basic clock rate to regain some lost performance.

The Bellmac 32 CPU (11) is another good example of structured design. This chip is interesting because it uses three different layout techniques in its design. These are standard cell (polycell), gate matrix and PLA for the control section (effectively a microprogram store). It also uses clever circuit techniques (ie, domino logic (22)) to improve speed while conserving power. The original chip fell short of the required performance, so a second version of the CPU was designed to increase throughput. The major contribution to the increased throughput seems to have been a modification of the internal architecture (mainly the control section) to reduce the average number of cycles per instruction (by a factor 2). This increased the complexity of the PLA control section from 65K sites to 100K sites. A reduced minimum feature size was also used for the 32A, which probably also aided performance, and allowed the increased amount of control logic to be integrated into a smaller area than the original - 1 sq.cm. @ 2.5 microns vs 1.5 sq.cm. @ 3.5 microns. It is also interesting to note that Bell decided that the 'polycell' section of the 32 would be better if implemented as gate matrix in the 32A - again, this could have been to improve performance, or possibly to be more consistent in design style.

As hierarchical, top-down design methodologies lead to a clear, comprehensible view of the system, they enable the real bottlenecks in system performance to be identified. This is true both of software as well as hardware, and the integration between them.

Smaller feature sizes are increasing throughput from basic improvements in devices speeds and reduced capacitance, and are allowing more tradeoff of silicon area for throughput, as seen by increased use of pipelining and cache/on-chip RAM.

VLSI DSP Architectures - An Overview

Single algorithm architectures can be found in many applications - digital signal processing (DSP), data processing, matrix processing, special purpose hardware simulators; we shall consider some basic DSP chip constraints.

The specification of a DSP function will be based upon the signal bandwidth (sampling rate) and the signal to noise ratio (word size). The signal bandwidth may well determine the technology, whilst the signal to noise ratio will determine whether fixed or floating point arithmetic is used. Common realisations of many DSP algorithms are based on multiply- accumulators with RAM and ROM functional blocks. These can in turn be combined to produce higher level primitives such as FFT butterflies and filter sections. Recently considerable interest has been shown in bit-level (systolic) algorithms for a variety of DSP functions (22, 23). Bit-level considerations also give rise to the bit-serial/parallel conflict. Note that bit-level approaches are not suitable for floating point data representations.

DSP functions are concerned mainly with number crunching in both the data path and address generation. Address sequences may be long and complex depending upon the exact algorithm and transform size, but are highly structured and may be either generated directly in hardware or via microcode. With the majority of DSP algorithms being free of branching processes, pipelined architectures may be implemented at the expense of a twofold increase in complexity, but with perhaps an order of magnitude increase in data rate. The resulting latency is in general only a problem in recursive algorithms.

Let us revisit the DFT. There are numerous algorithms and variants available to compute such a transform. They offer trade-offs in the number of multiplications, additions, memory and control (24). Some algorithms indeed have no true multiplications (25), or use coefficients that have a binary representation sparse in 1's, although at the expense of some accuracy. But which do we choose from technology considerations?

Advanced DSP functions such as the FFT are using increasingly large data words and rates. For a radix-2 rank pipelined FFT then a complex (4 real) multiplication per rank is required. For say a 512-point FFT, then a total multiplier complexity of up to 63K-140K gates is required depending on the depth of pipelining and assuming each real multiplier is multiplexed two ways. With say a further 20K equivalent gates for data storage, such a design becomes impractical on a single chip with present (2.5 micron, power-delay product = 0.16pJ) technology. With highly structured DSP algorithms then modularity to minimise the number of designs is a major requirement. By cascading modular chips the transform size or word size may be altered to suit. For (16+16) bit complex data, then a radix-2 FFT butterfly calls for 160 pins for data and coefficient only. The 64 output pins alone will dissipate some 0.4 watts at a data rate of 10MHz (25pF loading). This is a stringent requirement for any packaging technology and has been recognised within the VHSIC programme; pin-count has been seen to be approximately equal to the square root of the gate complexity.

An alternative is to multiplex pins, but this can lead to severe timing requirements on external chip communication. For CMOS technology, however, pin multiplexing does not represent a reduction in power dissipation in the output drivers. This appears to represent an all or nothing situation, with total system integration being the goal. Wafer scale integration may assist in this area (26). However, if the data rate can be reduced ie. low PRF radars or in audio bandwidth applications, then bit-serial arithmetic can be used to considerable advantage.

Consider a 5-pole IIR anti-aliasing filter for PCM applications. With 16 bit data and coefficients and an effective sampling rate of 32KHz, a bit rate of 10.24Mbits/s is required, well within VLSI CMOS performance of say, 20MHz. The conclusion is that bit-serial is alive and increasing in importance.

Multipliers are well known for their design and complexity problems. Like the FFT, there are many multiplier algorithms, Booths, Baugh-Wooley, carry-save etc. Parameterised parallel multipliers based on standard-cells are thus attractive but with a high area overhead. However, performance is limited by the way in which carries are propagated; the choice of algorithm is therefore highly technology dependent indicating a (automated) custom approach. Alternatively, a bit-serial approach is inherently easier to parameterise (27) as fan out/in is predictable. Further, for an N*N serial multiplication there is an approximate N-times reduction in complexity and power over the parallel case.

Standard-cell (including parameterised cells) is the only viable design methodology for single algorithm DSPs. The major demand is for parameterisable RAM and ROM (of relatively small size but possibly multiport and high speed - sub 50nS), adder, and multiplier cells. RAM and ROM cells are already available, with adders part of the bit-sliced ALU. In general performance is presently marginal with current technology.

DSP functions differ from general purpose MPU chips in that MPU chips are often memory dominated. DSP chips may be configured as (systolic type) arrays of arithmetic logic. It may be expected therefore that the total equivalent gate complexity achievable for DSP may be less than for a MPU, due to limitations on area (yield) and power.

Discussions and Conclusions

We have tried to show that design methodology is not just related to logic layout in silicon. Design methodology refers to the whole design from system concept to physical realisation. A universal design methodology has yet to emerge and perhaps is unlikely to emerge; each one particular methodology reaches a different compromise in performance, design cost, and production cost.

Whatever methodology is chosen we must retain the ability to introduce changes, not only in specifications, but also in technology, in design rules, and with error detection and correction. With this ability we can ease custom design requirements by introducing such synthesis tools as silicon compilers. Technology independent parameterisable cells attempt to satisfy these needs, but in general, only modify topology and functionality, not performance. Custom circuit design will remain for some time yet.

Single algorithm dedicated processors/architectures will demand very high complexity chips. Modularity, with a minimum number of designs is essential, but may be limited by communication (data) paths ie pin-out.

VLSI offers complexity. This advantage should be taken to enable total system requirements and costs to be reduced.

REFERENCES

1. Yuan H.T. et al, 1982, "Properties of Interconnection on Silicon, Sapphire, and Semi-Insulating Gallium Arsenide Substrates", IEEE Journal of Solid-State Circuits, SC-17, 269-274.

2. Saraswat K.C., et al, 1982, "Effect of Scaling of Interconnections on the Time Delay of VLSI Circuits", IEEE Journal of Solid-State Circuits, SC-17, 275-280.

3. Pritchard W.,1982, "A CMOS PLA Generator", Proc. ESSCIRC, 94-96.

4. Fillmore R., 1983, "Standard Cell Expandable Static CMOS RAM", Proc. CICC, 115-122.

5. Donze R. et al, 1982, "PHILO - A VLSI Design System" ACM IEEE 19th Design Auto. Conf.,163-169.

6. Budde D.L. et al, 1981, "The Execution Unit for the VLSI 432 General Data Processor", IEEE Journal of Solid-State Circuits, SC-16, 514-521.

7. Bayliss J.A. et al, 1981, "The Interface Processor for the Intel VLSI 432 32-bit Computer", IEEE Journal of Solid-State Circuits, SC-16, 522-530.

8. Bayliss J.A. et al, 1981, "The Instruction Decoding Unit for the VLSI 432 General Data Processor", IEEE Journal of Solid-State Circuits, SC-16, 531-537.

9. Slager J. et al, 1983, "A 16b Microprocessor with On-Chip Memory Protection", IEEE ISSCC Digest of Technical Papers, 24-25.

10. Ochester R., November 3 1981, "Low-cost 16-bit microprocessor has performance of midrange minicomputer", Electronics, 129-133.

11. Murphy B.T., 1983, "Microcomputers: Trends, Technologies and Design Strategies", IEEE Journal of Solid-State Circuits, SC-18, 236-244.

12. Mor S. et al, 1983, "A 16b Microprocessor for Realtime Applications", IEEE ISSCC Digest of Technical Papers, 28-29.

13. Alpert D. et al, July 14 1983, "32-bit Processor Chip Integrates Major System Functions", Electronics, 113-119.

14. Magar S.S. et al, 1982, "A Microcomputer with Digital Signal Processing Capability", IEEE ISSCC Digest of Technical Papers, 32-33.

15. Mead C. and Conway L., 1980, "Introduction to VLSI Systems", Addison-Wesley.

16. Hayn J. et al, January 27 1981, "Strip architecture fits microcomputer into less silicon", Electronics, 107-111

17. Guttag K.M. et al, 1982, "A 16b Microprocessor with 152b Wide Microcontrol Word", IEEE ISSCC Digest of Technical Papers, 120-121.

18. Beyers J.W. et al, 1981, "A 32-bit VLSI CPU Chip", IEEE Journal of Solid-State Circuits, SC-16, 537-542.

19. Patterson D.A. and Sequin C.H., 1981 "RISC 1: A reduced instruction set VLSI Computer", Proceedings of the Eighth International Symposium on Computer Architecture, 443-457.

20. Nakamura H. et al, 1983, "A Circuit Design Methodology for CMOS Microcomputer LSIs", IEEE ISSCC Digest of Technical Papers, 134-135.

21. Lineback J.R., December 15 1982, "Serial design halves processor chip size", Electronics, 48-49

22. Krambeck R.H. et al, 1982, "High Speed Compact Circuits with CMOS", IEEE Journal of Solid-State Circuits, SC-17, 614-619.

23. McCanny J.V., McWhirter J.G., 1982, "Implementation of Signal Processing using 1-bit Systolic Arrays", Elect. Letts., 18, 241-243.

24. Wood D. et al, 1983, "An 8 Bit Serial Convolver Based on a Bit Level Systolic Array", Proc. CICC, 256-261

25. Blanken J.D. and Rustan P.L., 1982, "Selection Criteria for Efficient Implementation of FFT Algorithms", IEEE Trans. Acoust., Speech, Sig. Proc., ASSP-25, 107-109.

26. Curtis T.E., Wickenden J.T., 1983, "Hardware Based Fourier Transforms: Algorithms and Architectures", IEE Proc., 130, Pt.F, 423-432.

27. Garverick S.L. and Pierce E.A., 1983, "A Single Wafer 16-Point 16-MHz FFT Processor", Proc. CICC, 244-248.

28. Denyer P.B. and Renshaw D., 1983, "Case Studies in VLSI Signal Processing Using a Silicon Compiler", Proc. ICASSP, 939-942.

FACTORS LIMITING PERFORMANCE IN MINIMUM GEOMETRY DEVICES AND CIRCUITS

G. Declerck

ESAT Laboratory, K.U.Leuven, Belgium

INTRODUCTION

The scaling laws for MOS transistors will first be reviewed and the most important device limitations will then be discussed in greater detail. Finally some considerations about speed and power consumption of scaled devices will be made.

SCALING LAWS

Device scaling has been extensively pursued during the last decade as a means to enhance circuit performance (higher speed and lower power consumption) and to increase packing density. According to the original constant-field scaling law, proposed by Dennard et al. (1), all geometries, the supply voltage and the device current are reduced by the scaling factor S, while the substrate doping is increased by 1/S. This results theoretically in a reduction of the gate delay by S and of the power dissipation by S^2. In practice however some parameters such as interconnection capacitances and parasitic resistances do not scale linearly with dimension and parameters as the subthreshold current-voltage slope and the built-in voltages do not scale at all. In order to preserve the compatibility with other logic families and also for considerations of noise margin, Chatterjee et al. (2) propose a non-constant field scaling in which the operating voltage is kept constant (constant voltage scaling), or is reduced at a slower rate (quasi-constant voltage scaling). The different scaling laws are summarized in table 1.

Several attemps have been made to estimate the performance of devices scaled under the various scaling laws. Chatterjee et al. (2) predicted an optimum performance for non-constant field scaling. In their study the available drive current showed a maximum at L = 0.4 - 0.5 µm for n-channel and at L = 0.3 µm for p-channel devices. A further scaling of p-channel devices seems to be possible, mainly because velocity saturation effects are less pronounced for holes ($E_c = 2 \times 10^5$ V/cm) than for electrons ($E_c = 2 \times 10^4$ V/cm). The junction sheet resistances were modelled as being inversely proportional to the junction depth x_j. Shichijo (3) and Scott et al. (4) have used a transmission line model for the parasitic resistances of the source and drain regions. In this model the sheet resistances of the shallow junctions are increasing much more rapidly than $1/x_j$ mainly because of the decrease of surface concentration. The results of this analysis are shown in fig.1, where the triode gain of n-channel and p-channel devices is plotted as a function of gate length. In the QCV-scaling, which is believed to most closely represent the current trend in industry, the gain for both n-channel and p-channel devices start to drop below 0.8 µm. This is mainly due to the combined effects of mobility degradation and parasitic resistances.

Other approaches of device scaling may shift the peak of device performance towards dimensions of 0.3 - 0.5 µm, however a consensus exists that below these dimensions only little device improvement will be achievable and that further scaling will mainly aim at higher functional density. Some of the device limitations will now be discussed in more detail.

DEVICE LIMITATIONS

Effects on threshold voltage. It is well understood that the threshold voltage V_T decreases when the effective channel length is reduced to a value comparable to the depletion layer width at source and drain. This phenomenon has been explained by Yau (5) as the Charge Sharing Effect. A proper application of this principle allows an accurate modelling of the V_T-dependence on channel length and on drain-source voltage and a precise prediction of the backgate bias effect can also be made (Mol and Sun (6) and Chatterjee and Leiss (7)).

A scaling of the channel width to one micron or less also affects the device threshold and severely reduces the effective channel width due to the lateral encroachment of the field region. The lateral extension of the bird's beak in the classical local oxidation technique can no longer be tolerated and improved isolation schemes are being studied at several laboratories.

TABLE 1 - Definition of the scaling laws.

Scaling Law	Constant Field CE	Constant Voltage CV	Quasi-Constant Voltage QCV
Dimensions	S	S	S
Gate oxide	S	$\sqrt{S}$ or S	S
Doping	1/S	1/S	1/S
Voltage (λ_V)	S	1	$S < \lambda_V < 1$
			e.q. $\lambda_V = \sqrt{S}$

Mobility degradation. At smaller effective channel lengths the electric field along the channel increases and the electron velocity deviates from its linear dependence on the field and finally reaches a scattering limited value v_{max} of about 1×10^7 cm/sec at fields above 5×10^4 V/cm. Holes reach approximately the same velocity saturation value at somewhat higher fields. In short channel MOS transistors the drain saturation voltage V_{DSAT} is no longer given by conditions of pinch-off at the drain but is now defined by the drain voltage at which the carriers reach the saturation velocity near the drain.

Besides its effects on the V_{DSAT}, the velocity saturation effect will also strongly reduce the drain current in both linear and saturation mode of operation. This can be expressed in a simple way by introducing an effective mobility μ_e as proposed by Hoeneisen and Mead (8)

$$\mu_e = \frac{\mu_o}{1 + \frac{E}{E_c}} = \mu_o F_v \quad (1)$$

μ_o is the low field mobility, E is the average field along the channel and F_v is the velocity saturation factor. The drain current can now be written as (El-Mansy (9))

$$I_D = F_v \cdot I_D (v_{max} = \infty) \quad (2)$$

The factor F_v is given in table 2 for devices with effective channel lengths ranging between 2.0 μm and 0.5 μm (9).

TABLE 2 - F_v as a function of L_{eff}. The average field is calculated at V_{DS} = 2.5V (taken from ref.9).

L_{eff} (μm)	2.0	1.5	1.0	0.5
F_v	0.55	0.46	0.36	0.22

It should also be remarked that in the limit, for very small L_{eff}, the saturation drain current will be completely dominated by carrier velocity saturation and will be given by:

$$I_{DSAT} = W C_{ox} (V_{GS} - V_T) \cdot v_{max} \quad (3)$$

The drain current no longer depends on channel length as the carriers travel at maximum velocity; the current is now linearly dependent on the gate voltage whereas the transconductance becomes independent of the gate voltage; the gate to drain and gate to source capacitances no longer fit the classical models (6).

A number of other physical effects start to play an important role for effective channel lengths smaller than 2.0 μm. Source and drain sheet resistances go up as junctions become shallower and contact resistances are getting very important when contact openings are reduced to dimensions of 1.0 μm or less. It should also be reminded that the carrier mobility itself is reduced by vertical electric fields and by the increased channel doping as shown by Sun and Plummer (10).

As the gate insulator thickness is reduced to 10 nm or less it is becoming comparable to the finite thickness of the inversion layer and this effect can be modelled by putting the inversion layer capacitance in series with the gate insulator capacitance. This results in a current reduction factor F_{CH} which is 0.86 for a device with a 20 nm thick gate oxide and 0.75 for a 10 nm thick oxide (9).

The effects of velocity saturation, of parasitic resistances and of finite inversion layer thickness have been brought together by El-Mansy (9) and are shown in fig.2. For these n-channel devices the velocity saturation effect is the most dominant gain reducing factor down to 0.5 μm channel length. In p-channel devices the main limiting factor will probably be the much higher sheet and contact resistance of shallow p^+-junctions (3).

Maximum operating voltage. The maximum operating voltage of a short channel transistor is limited by several mechanisms : source to drain punchthrough, parasitic bipolar breakdown, gate insulator breakdown and hot carrier related phenomena.

Punchthrough occurs when the potential barrier at the source is lowered by field lines extending from the drain depletion layer. This drain induced barrier lowering results in an enhanced current injection from the source into the channel region (Kotani and Kawazu (11), Barnes et al. (12), Eitan and Frohman-Bentchkowski (13)). Although the conditions for the channel implants are optimized in order to minimize the punchthrough currents, punchthrough is considered as one of the most serious voltage limitations for devices scaled below 0.5 μm gate length.

Another fundamental limitation to the maximum drain voltage of n-channel devices is imposed by the parasitic bipolar breakdown, which is also called the snapback effect ((13), Sun et al. (14), Toyabe et al. (15) and Toyabe and Asai (16)). The mechanism is triggered by impact ionization near the drain, resulting in a flow of majority carriers towards the substrate. This substrate current causes a forward biasing of the source to substrate junction, which turns on the parasitic npn bipolar transistor and reduces the sustaining voltage of the MOS transistor. The use of graded junctions or lightly doped drains has been shown to raise the sustaining voltage over a few volts by decreasing the multiplication effects near the drain (Tsang et al. (17) and Sunami et al. (18)).

The operating voltage of scaled devices is severely limited by the presence of several hot carrier related mechanisms which eventually lead to long term reliability problems (Ning et al. (19), Takeda et al. (20) and Takeda et al. (21)). Carriers travelling from source to drain in a short channel transistor can gain enough energy to surmount the Si-SiO$_2$ energy barrier and to get injected into the gate insulator. Trapping of a fraction of these injected carriers in the gate dielectric material, or creation of interface traps gives a threshold voltage shift and a reduction of the transconductance. Electron and hole injection can also be initiated by weak avalanching near the drain. The substrate currents, originating from the same impact ionization, lead to the snapback breakdown effect, to latch-up problems in CMOS and to non-thermal leakage current problems in dynamic memory structures or CCD's due to secondary impact ionization.

The sustaining voltage due to the bipolar snap back effect is plotted in fig.3 as a function of the effective channel length. The same figure also shows the highest applicable voltage as imposed by hot carrier reliability considerations (20). For devices with an effective channel length of 0.5 μm the maximum voltage drops to 3.5V. Optimization of the drain doping profile reduces the drain electric field and allows operation at slightly higher voltages. However it is clear that the various hot carrier phenomena deserve great attention, especially if the operating voltages are not scaled linearly with geometries.

PERFORMANCE OF SCALED TECHNOLOGIES

Making a quantitative comparison between different scaled technologies is difficult as no standardized tools exist to study speed, density and power consumption. The commonly used ring oscillators are composed of different gates (inverters, NOR's or NAND's) with different loading conditions and with different aspect ratio's for the transistors. It should also be emphasized that the performance of VLSI-circuits will to a great extent be dominated by the parasitics of the interconnects and not by the optimum performance of the basic gates.

MOS technologies with 1 μm gate lengths will give us power-delay products (Pt_d) between 0.02 and 0.4 pJ with minimum gate delays of less than 0.5 nsec. Propagation delays of 0.1 to 0.2 nsec have been reported by several laboratories for 1 μm CMOS processes (Sakai et al. (22) and Yamaguchi et al. (23)) and 1.5 μm nMOS (Liu et al. (24)) respectively.

When comparing nMOS versus CMOS it is recognized that for the same geometries both technologies can have the same speed, but CMOS has the big advantage of its very low power consumption. While its static power consumption is almost negligible, its dynamic power consumption is also 5 to 10 times smaller than for nMOS. This has been clearly demonstrated by comparing a 4k static RAM fabricated in HMOS II (with Pt_d = 0.2 pJ) and in HMOS - CMOS (with Pt_d = 0.04 pJ). (Yu et al. (25)). It should of course be reminded that when part of a CMOS VLSI circuit operates at a lower frequency, the power consumption is proportionally reduced.

It is also generally accepted that for the same geometries bulk CMOS behaves about twice as slow and consumes twice as much power as its SOS/CMOS counterpart (White (26)). However it is still believed that in the near future bulk CMOS will form the mainstream VLSI process while non-bulk CMOS processes will become very attractive in the late 80's as soon as cheaper silicon-on-insulator techniques will become established.

REFERENCES

1. Dennard, R.H., Gaensslen, F.H., Yu, H., Rideout, V.L., Bassous, E., and LeBlanc, A.R., 1974, IEEE Journ. of Solid-State Circ., 9, 256-268.

2. Chatterjee, P.K., Hunter, W.R., Holloway, I.C., and Lin, Y.T., 1980, IEEE Electron Device Letters, 1, 220-223.

3. Shichijo, H., 1981, IEDM Tech. Digest, 219-222.

4. Scott, D.B., Hunter, W.R., and Shichijo, H., 1982, IEEE Trans. on Electron Devices, 29, 651-661.

5. Yau, L.D., 1974, Solid-State Electronics, 17, 1059-1063.

6. Moll, J.L., and Sun, E.Y., 1980, Jap. Journal of Appl. Phys. Supplement, 19-1, 77-83.

7. Chatterjee, P.K., and Leiss, J.E., 1980, IEDM Techn. Digest, 28-33.

8. Hoeneisen, B., and Mead, C.A., 1972, IEEE Trans. on El. Dev., 19, 382-383.

9. El-Mansy, Y., 1982, IEEE Trans. on El. Dev., 29, 567-573.

10. Sun, S.C., and Plummer, J.D., 1980, IEEE Journ. of Solid-State Circ., 15, 562-573

11. Kotani, N., and Kawazu, S., 1979, Solid-State Electronics, 22, 63-70.

12. Barnes, J.J., Shimohigashi, K., and Dutton, R.W., 1979, IEEE Journ. of Solid-State Circ. 14, 368-375.

13. Eitan, B., and Frohman-Bentchkowski, D., 1982, IEEE Trans. on Electron Devices, 29, 254-266.

14. Sun, E., Moll, J., Berger, J., and Alders, B., 1978, IEDM Tech. Digest, 478-482.

15. Toyabe, T., Yamaguchi, K., Asai, S., and Mock, M., 1978, IEEE Trans. on Electron Devices, 25, 825-832.

16. Toyabe, T., and Asai, S., 1979, IEEE Journal of Solid-State Circ., 14, 375-383.

17. Tsang, P.J., Ogura, S., Walker, W.W., Shepard, J.F., and Critchlow, D.L., 1982, IEEE Trans. on Electron Devices, 29, 590-596.

18. Sunami, H., Shimohigashi, K., and Hashimoto, N., 1982, IEEE Trans. on Electron Devices, 29, 607-610.

19. Ning, T.H. Cook, P.W., Dennard, R.H., Osburn, C.M., Schuster, S.E., and Yu, H., 1979, IEEE Trans. on Electron Devices, 26, 346-352.

20. Takeda, E., Kume, H., Toyabe, T., and Asai, S., 1982, IEEE Trans. on Electron Devices, 29, 611-618.

21. Takeda, E., Nakagome, Y., Kume, H., and Asai, S., 1983, IEE Proc., 130, 144-149.

22. Sakai, I., Kudoh, O., and Yamamoto, H., 1982, IEDM Techn. Digest, 702-705.

23. Yamaguchi, T., Morimoto, S., Kawamoto, G., and DeLacy, J., 1983, Proc. of the Custom Int. Circ. Conf., 57-60.

24. Liu, S. Fu, C., Atwood, G., Dun, H., Langston, J., Hazani, E., So, E., Sacholev, S., and Fuchs, K., 1982, IEEE Journ. of Solid-State Circuits, 17, 810-814.

25. Yu, K., Chwang, R., Bohr, M., Warkentin, P., Stern, S., and Berglund, N., 1981, IEEE Journ. of Solid-State Circuits, 16, 454-459.

26. White, M., 1982, IEEE Trans. on Electron Devices, 29, 578-584.

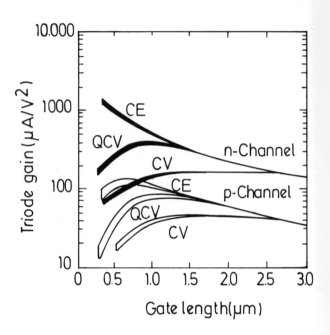

Figure 1 Triode gain of scaled MOSFET's versus patterned gate length for n-channel and p-channel devices assuming different scaling laws (After Scott et al, reference 4)

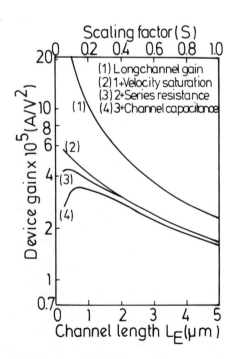

Figure 2 The degradation of device gain due to velocity saturation, series resistance and channel capacitance plotted versus channel length (After El-Mansy, reference 9)

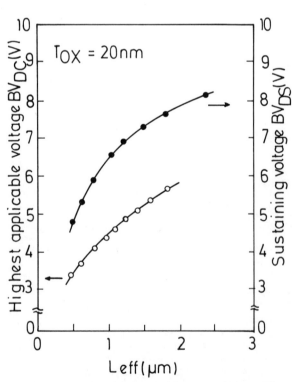

Figure 3 The highest applicable voltage BV_{DC} and the sustaining voltage BV_{DS} as a function of channel length (after Takeda et al., reference 20)

THE TECHNOLOGY CHOICE: CMOS VS NMOS VS BIPOLAR

D.L. GRUNDY

FERRANTI ELECTRONICS LIMITED

INTRODUCTION

This paper sets out to compare the relative merits of NMOS, CMOS and bipolar technologies. In previous comparisons emphasis has been placed upon performance factors such as speed, power dissipation and packing density. Whilst these are obviously important qualities; with increasing maturity of the semiconductor industry it is becoming increasingly apparent that the economics of manufacture are of increasing concern and with this paper it is intended to cover some aspects of manufacturing cost.

NMOS

The LSI business started with PMOS in the late sixties, PMOS came first because it was found easier to establish a controllable channel in the presence of contaminating charge trapped in the gate oxide region when compared with NMOS. Unfortunately PMOS was relatively slow certainly when compared with bipolar of the day and there was considerable incentive to switch to NMOS where the threefold improvement in mobility of electrons compared with holes promised a corresponding increase in speed. The NMOS technology arrived when greater control over oxide quality was established and in addition the benefits of the evolving ion implanting art could be applied.

NMOS presently has no rival for memory products such as ROM and RAM, industry standards of 64K and the emerging 256K are all manufactured this way and look unlikely to change. The main advantage of NMOS is high packing density, a basic memory bit comprising just one transistor and overlapped capacitor for dynamic and two transistors with polysilicon loads for static. Other impressive products include the Hewlett Packard (Mikkelson et al (1)) 32 bit computer with a functional throughput rate of 6×10^{12} gate Hertz per square centimetre.

A major disadvantage of NMOS is power dissipation. The speed power product E_{SP} is given approximately by:

$$E_{SP} = \frac{V_{CC} V_T C_L}{2} \quad \ldots \ldots \ldots \ldots \ldots \ldots (1)$$

V_{CC} = supply volts, V_T = threshold voltage, C_L = load capacitance with 5 volts supply and 1 volt thresholds this gives 2.5 picojoules for a 1pF load capacitor. This means that a fast complex chip involving thousands of gates could require many watts of power.

A further problem is limited off chip drive capability, high capacitive loads and transmission lines ideally require "ON" resistance values measured in tens of ohms for the source and sink output modes which are difficult to achieve with a basic sheet resistivity of 10K per square.

Finally NMOS is very limited in terms of linear performance.

CMOS

From a digital design point of view CMOS is very attractive, mutually exclusive conduction modes enable near zero quiescent dissipation if leakage is under control. Early forms of CMOS did not compare favourably with PMOS or NMOS in terms of packing density due to the requirement for a separate isolation region for the N channel device and also for a greater number of transistors to form the basic gate. Introduction of silicon gates, ion implantation, finer geometries and latterly oxide isolation have all contributed to the establishment of a modern day highly competitive form of CMOS with many performance advantages over NMOS if not necessarily providing the ultimate in packing density.

The main advantage is reduced power dissipation, since there is no direct path to ground for supply current dissipation is almost entirely dynamic given by:-

$$P = f C_L V_{CC}^2 \quad \ldots \ldots \ldots \ldots \ldots \ldots \ldots \ldots (2)$$

This is a maximum value, in a practical logic design comprising hundreds of gates only a percentage undergoes a transition on each clock cycle, the magnitude of this percentage is very much dependent on the type of system and lies between 5% and 50%. The former figure applying to something like a long binary counter whilst the latter would apply to worst case data patterns in a long shift register. A typical average is 10% inserting into (2) gives:-

$$P = 0.1 f C_L V_{CC}^2 \quad \ldots \ldots \ldots \ldots \ldots \ldots \ldots (3)$$

This results in very low dissipation at low clock rates but can approach one watt for complex high speed circuits.

The off chip drive capability of CMOS is superior to NMOS due to availability of low resistance devices in source and and sink modes but there are still difficulties with the driving of highly capacitive loads and transmission lines. In addition CMOS offers the linear designer more degrees of freedom than NMOS and there is a growing tendency to include linear functions on CMOS chips.

A main disadvantage of junction isolated CMOS is a tendency to latch up. This occurs when a large uncontrollable supply current flows as a result of voltage spikes on supply or input/output pins or in some cases within the chip core when fast edges are capacitively coupled into floating nodes.

BIPOLAR

Bipolar technology predates all forms of MOST by a period approaching five years. The switch to MOST occurred due to the problems associated with yielding LSI chips on standard buried layer bipolar and also to the excessive power consumption demands of certain bipolar logic families.

If ultimate speed is the goal bipolar has much to offer. The conducting channel once established has remarkably low resistance, when taken in conjunction with the capacitive loads in the 1pF order time constants in the tens of picosecond region are possible.

Switching delay in bipolar is defined ultimately by transit time across the base given by:-

$$t_B = \frac{W^2}{2.43 \, D_N \, I_N \, \frac{N_E}{N_C}} \quad \ldots \ldots \ldots (4)$$

W = base length, D_N = diffusion constant, N_E = impurity concentration at emitter, N_C = impurity concentration at collector.

The base length W is seen to be critical and for sub-micron values which bipolar routinely achieves transit times less than 100 picoseconds are possible. With further advances in bipolar technology such as poly-silicon emitters it is easy to imagine that the sub 100 picosecond gate is not very far away.

From a historical point of view bipolar is always considered to be excessive in its power consumption requirements. This may have been true for early ECL but not for modern bipolar technology such as Ferranti FAB-2 (Grundy and Bruchez (2)). The reason for this is that bipolar is able to operate with small logic swings certainly less than 100 millivolts, speed-power product is given by:-

$$E_{SP} = V_{CC} \, V_L \, C_L \ldots \ldots \ldots (5)$$

V_{CC} = supply voltage, V_L = logic swing, C_L = load capacitance. With a 1 volt supply, 100 millivolts logic swing and load of 1pF the speed power product is 10^{-13} picojoules and this can be reduced even further.

In addition to extremely high speed and low power dissipation bipolar has no problems with off chip drive capability offering up to 100mA in either source or sink modes.

Finally and very significantly the well proven linear capability of bipolar will prove to be a vital asset for VLSI applications.

On the negative side it is difficult presently for bipolar technologies to achieve a quiescent current consumption which approaches that of CMOS. In addition the zero offset capability for analogue switches associated with MOST technology is not achievable and neither is the zero input current of amplifiers.

Performance Comparison

For similar products such as gate arrays using similar geometries and design philosophies a quick scrutiny of manufacturers published data reveals that there is perhaps little to choose in terms of packing density for the three technologies and this will not be debated further.

Turning to switching speeds there are presently much larger differences and an attempt to summarise these is made in figure 1. This is a plot of switching time as a function of the critical conducting path dimension which for MOST technology is the channel length and for bipolar is the base length. Equation (4) is plotted for bipolar which shows a base transit time of 150 picoseconds for a 1 micron base length which is typical of Ferranti FAB-2(2). Interestingly the actual fastest gate delay for the R series based on FAB-2(2) is presently 1 nanosecond which indicates that there is scope for considerable improvements by reducing base resistance, load capacitance etc.

The switching delay for MOST is plotted with load capacitance as a parameter. With todays 3 micron channel length and assuming a $\frac{W}{l}$ ratio of 10 to 1 the basic time constant for the slower p channel device will be approximately 2 nanoseconds with a 1pF load.

Whilst scaling of the MOST transistor will obviously reduce both channel resistance and load capacitance since the latter must include interconnection factors the analysis is simplified if this is treated as a parameter.

As both technologies evolve it is apparent from Figure 1 that in theory similar speeds are achievable, if gate delays of 100 picoseconds are required however submicron MOST channels become necessary which will be difficult if not impossible to manufacture reliably.

Next consider power dissipation:

For NMOS it has been shown that the speed power product is given by equation (1) and the corresponding power is derived by dividing by the gate delay t_D hence:-

$$\text{Power Dissipation NMOS} = \frac{V_{CC} \, V_L \, C_L}{2 t_D} \ldots \ldots (6)$$

and similarly for bipolar:

$$\text{Power Dissipation BIPOLAR} = \frac{V_{CC} \, V_L \, C_L}{t_D} \ldots \ldots (7)$$

The form of the corresponding equation for CMOS is rather different as given by (3):

For purposes of comparison it is necessary to express (6) and (7) in terms of frequency which is achieved by assuming that the allowable gate delay t_D must not exceed one fifth of a clock period:-

$$t_D = \frac{0.2}{f_c} \ldots \ldots \ldots (8)$$

The following assumptions are then made:

C_L = 1pF, V_L = 0.25 volts, V_{CC} = 5 volts, V_T = 1 volt.

This gives:

$$\frac{NMOS}{CMOS} \text{ power} = \frac{2.5 \, V_{CC} \, V_T \, C_L \, f}{0.1 \, f \, V_{CC}^2 \, C_L} = 5:1 \quad \ldots\ldots(9)$$

$$\frac{NMOS}{BIPOLAR} \text{ power} = \frac{2.5 \, V_{CC} \, V_T \, C_L \, f}{5 \, V_{CC} \, V_L \, C_L \, f}$$

2:1 (V_{CC} BIPOLAR = 5 volts),

5:1 (V_{CC} BIPOLAR = 2.0 volts) ....(10)

$$\frac{CMOS}{BIPOLAR} \text{ power} = \frac{0.1 \, f \, V_{CC}^2 \, C_L}{5 \, V_{CC} \, V_L \, C_L \, f}$$

0.4:1 (V_{CC} BIPOLAR = 5 volts)

1:1 (V_{CC} BIPOLAR = 2 volts) ....(11)

Finally a complete comparison with star ratings is shown in Table 1 for all important aspects of performance for the three technologies.

TABLE 1 - Performance Comparison

		NMOS	CMOS	BIPOLAR (FAB2)
D I G I T A L	SPEED	***	***	****
	POWER	**	*****	****
	COMPLEXITY	*****	****	****
	LATCH UP	N	Y	N
	DRIVE CAPABILITY	**	**	****
		NMOS	CMOS	BIPOLAR (FAB2)
L I N E A R	HIGH FREQUENCY	*	**	****
	DRIFT	*	*	****
	VOLTAGE REFERENCES	**	***	*****
	ANALOGUE SWITCH	*****	*****	***
	NOISE	**	**	***
	AMPLIFIER INPUT CURRENT	*****	*****	**

MANUFACTURING ECONOMICS

Probably the most significant factor related to manufacturing economics is the total number of masks involved in the processing sequence. This affects yields for both mask making and processing in addition to the basic capital investment required.

The composite yield for a set of masks is given by:-

$$\% \text{ yield} = 100 - DML^2 \left(\frac{d}{1+d}\right) 10^{-4} \quad \ldots\ldots(12)$$

D = significant pinhole density, M = number of masks per device, L = device size (mils), 1:d = clear to opaque ratio. This expression relates to mask yield and defines just how many good photographic sites will be defined When operating with mask sets greater than 10 or 12 with chip sizes in the region of 0.4 inches and a clear to opaque ratio of 50% a mask yield of 90% will require a defect density of 0.125 per square inch which is extremely low and both difficult and costly to achieve. This is the type of situation faced by many MOST manufacturers and for complex versions of bipolar. There is obviously a very great incentive to opt for fewer masks and the total of 6 required for the Ferranti FAB-2(2) process eases the mask making problem considerably allowing twice the defect density for similar yield.

When consideration is given to the actual cost of processing the need for a low mask count is emphasised even further. A modern projection aligner costs around £0.5 x 10 and the total number required is obviously proportional to the number of masks in the sequence. In a typical production unit operating 3 shifts and processing 2000 wafers per week on a 6 mask process such as FAB-2(2) there would be a requirement for 4 aligners at a cost of £2.0 x 10^6 and for 12 masks the same production rate would require 8 at a cost of £4.0 x 10^6. A further important aspect of photolithography is the capability of these machines in respect of minimum feature size. In general the more masks in a sequence the greater the difficulty factor associated with achieving a given minimum feature size, some impression for this can be gained from figure 3 which compares a 10 mask CMOS process with FAB-2(2). The flatter surface profile for the latter is of considerable advantage in easing the depth of field requirements for the aligner optics. The nett effect of this is that an aligner capable of resolving 2 microns with 6 masks may prove to be capable of only 3/4 microns with 10/12 masks. This is very important since it defines the point at which the switch to more exotic lithographic techniques such as DSW must occur. If for instance in the previous example a 2 micron feature size was required with 12 masks there would be little option but to use DSW machines costing £0.7 x 10^6 each and 12 such machines would be required totalling £8.4 x 10^6 which is a staggering 4.2 times larger investment.

In addition to reduced capital investment a lower mask count also means lower costs in direct proportion. In simple terms if we assume similar yields and area a chip requiring 12 masks will cost twice as much as one with 6 masks.

Conclusions

Whilst CMOS and bipolar technologies are beginning to offer overlapping performance in terms of speed and power dissipation there is evidence that the manufacturing economics associated with simplified bipolar processes may well prove to be a decisive factor for the future.

REFERENCES

1. Mikkelson, J.M., et al, An NMOS VLSI process for fabrication of a 32 bit chip, *IEEE Journal of Solid State Circuits* Vol. SC-16, No. 5, October 1981.

2. Grundy, D.L., and Bruchez, J.A., Switching to bipolar technology for the coming 100,000 gate arrays, *Electronics International*, Vol. 56, No. 14, July 14, 1983.

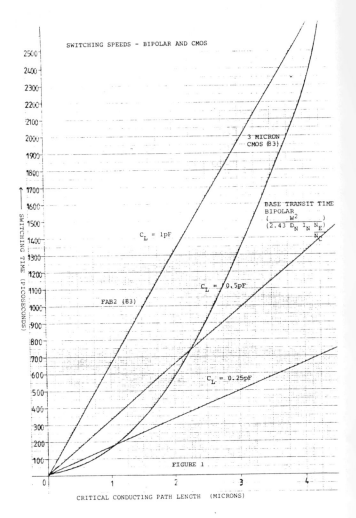

FIGURE 1

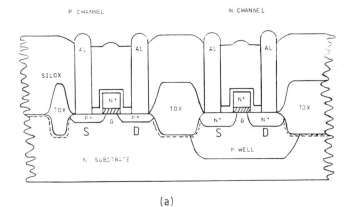

(a)

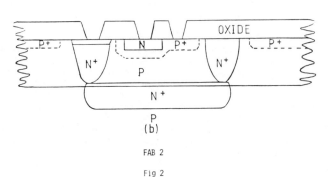

(b)

FAB 2

Fig 2

CHAIRMEN'S REPORT - SESSION 5 "SYSTEM AND PRODUCT APPLICATION AREAS"

H.J. Newman* and W. Fawcett**

*IBM UK Laboratories, UK, **Alvey Directorate, Department of Trade and Industry, UK

The application of leading edge/state-of-the-art VLSI in most commercial products today is mainly limited to microprocessors, their immediate support chips and memory. However, custom and semi-custom parts usage are projected to grow rapidly so that by the mid-80s they will account for half the merchant shipments. Since most of these custom parts will be 'customer' designed, the need for improved design systems with user friendly interfaces was stressed by many speakers - as was the need for re-education of product designers to exploit the potential of VLSI.

By 1990, with half the semiconductor consumption being application specific designs, it is clear that custom designs are perceived as providing product designers with a competitive edge - but only if these complex parts can be designed, fabricated and integrated into products quickly. This puts a premium on error-free design and fast path IC manufacture.

Telecommunications offers a significant VLSI opportunity, since many of the functions in the total network are repetitive. The challenge of mapping many traditional analogue functions into 'Analogic' ICs remains, but the rewards for companies who solve the problems are clear in terms of cost, function, reliability and equipment floor space.

Analysts' projections show 'Personal Computers' as a major market for VLSI products. In addition to the 16 and 32 bit minis, the need for gate and cell arrays is clear to provide 'glue' logic and unique 'value added' function. Existing PC manufacturers, who largely established the business, will experience mounting competition from vertically integrated companies who have advantages in 'control' of the total design, procure, build, sales and service of the product. This competitive environment will lead to highly innovative designs and short product life. The availability of engine architecture compatible software is perceived as a natural outgrowth of this growing opportunity.

Consumer electronics will continue to grow equally fast, although the specific applications and their share of the growth is not too clear. The vast opportunities in 'home' products and the IC volumes associated with it, make marketing in this area hazardous for any IC vendor since price competition is exceedingly strong - as demonstrated by the $2 watches available today. Clearly this is a volatile market.

Authors of papers on Military Applications all identified applications for VLSI and the need to apply state-of-the-art techniques in a similar manner to the commercial world. High data throughput is key in many military systems and hence custom designs are more effective than microprocessor solutions. However, low volumes per part number imply the need for low cost development and this can only be achieved with the aid of comprehensive tools to aid designers to design 15-30,000 gate chips quickly and 'right first time'. To achieve this, the multiple use of pre-tested macros in a standard cell approach or multiple application chip sets for a range of equipment designs was suggested.

As chips grow in complexity and size, the need for chip carriers with 130-200 + I/Os was highlighted. This package development for UK military systems needs attention and would benefit from a unified approach.

AN OVERVIEW OF EMERGING MARKETS AND REQUIREMENTS FOR CUSTOM/SEMICUSTOM VLSI IN THE COMMERCIAL MARKET

Malcolm G. Penn

Dataquest, UK

The industry shift towards customized semiconductor technology is profound and will have as significant an impact on the world electronics industries as did the microprocessor revolution that occurred in the 1970's. I would like to talk today on such "non-standard" integrated circuits, that is "Application-Specific Integrated Circuits".

The number of different approaches to Application-Specific Integrated Circuits (ASICS) is rapidly increasing. Gate arrays are vying with standard cell custom circuits; programmable logic arrays are filling the low gate count applications; alterable microprocessors are becoming popular and each year more and more engineers are being trained in the Mead-Conway approach to structured logic design. Whilst the emphasis of my talk today will be on VLSI we feel very strongly that gate arrays, custom integrated circuits, standard cell integrated circuits, PAL devices, PLAs, and Mead-Conway chips are all part of the same market trend. This trend is towards circuits that are designed for a single application, and that can only be supplied to one customer-circuits that are "application-specific".

In this talk I'll be discussing the reasons for the swing to these application-specific integrated circuits, and sometimes I'll abbreviate the term to "ASIC". Following the discussion of the reasons for the trend, I'll give some additional evidence that supports the fact that such a trend does indeed exist. Following this, I'll discuss the impact of the ASIC trend on semiconductor companies, and then the impact on semiconductor users. Finally, I'll try to cover what I think still needs to be done.

As it says in the slide, we first put flip-flops on a chip, and then counters, and then minicomputers (of course, the minicomputer on a chip caused the microprocessor revolution), and then whole computers. The technology is now expanding further and chips are still getting more complicated, so soon we'll have a triple-computer chip. The problem is that none us ever heard of a standard triple computer. This is the first reason for the trend to ASICs; namely, that LSI circuits are becoming as complex as the systems they replace. Since few systems of this complexity have been built, the standards do not exist at the system level that can be converted to standards at the chip level. In other words, the semiconductor industry has run out of system architectures to copy to chips. As a result, the new chips will all have new architectures. We believe it will be some time before significant standards emerge.

These are the forces at work. First, there are few standard VSLI circuits; and second, these ASICs offer significant cost savings to the system manufacturer. Finally, ASICs can also offer significant speed advantages.

This slide given an example of the kinds of cost savings that can be achieved. I've taken a 20,000 gate system and compared the total cost for 10,000 units, assuming it is built with standard parts, digital semi-custom parts, or customer-designed VLSI devices. In ths example, it is assumed that a standard cell approach is used for the customer-designed devices. As you can see, the number of parts in the system shrinks dramatically from 33 parts to 2 in the gate array case, and to 1 part in the customer-designed case. The savings in both cases is well over 70 percent of the cost, even when the extra tooling cost is taken into account.

The next slide shows a microprocessor example. This is a garden variety kind of an example. It has one CPU, two support chips, and a 64K ROM. Notice that if it takes three lines of code to equal a gate, than a 64K ROM is equivalent to around 2,600 gates. However, these 2,600 gates must apply to all four chips. So the equivalent gate count per chip is around 666 gates. It's very interesting to notice that this is about the gate count of the gate arrays currently being shipped. Most gate arrays are being shipped in the 500 to 1,000 gate range.

Continuing the microprocessor example a little bit further, notice that such a project requires 8,000 lines of code. If it were coded in machine language at a cost of $25 per line, the development cost would be $200,000, certainly enough money to fund the development of four fully custom chips. If it were coded in a high-level language at one-tenth the cost per line of machine code, then the total development cost for programming the chips would be around $20,000. This is comparable to the cost of coding a gate array with 1,500 to 2,000 gates.

Continuing the example a bit further, let's assume that the microprocessor under question has a 100-nanosecond cycle time; therefore, if it takes three lines of code to be equivalent to one gate, it will take at least 300 nanoseconds to do what a gate does. If one could magically design a gate array to perform the same function, such an array could easily work ten times as fast at the gate level. In addition, all gates work simultaneously rather than sequentially. It's quite possible that the 1,000 gates would work in parallel rather than in series. Therefore, the speed advantage might approach as much as 10,000 to 1 over the microprocessor example.

What is the evidence supporting the market growth? The two facts that we are aware of are that the number of captive manufacturers is increasing substantially, and that there are many new ASIC start-ups. This slide shows a comparison of the merchant manufacturers and captive manufacturers. It can be seen that the captive manufacturer count has, in fact, increased significantly. We believe that captive manufacturers have been building factories because they have been unable to obtain ASICs from the merchant market. They have not been building factories to enter the RAM business, and certainly, that's a very wise decision on their part.

The next slide shows the distribution of semiconductor start-ups since 1977, and you can see that fully 52 percent of these start-ups are firms offering some kind of application-specific integrated circuit.

Well, what will the impact of ASICs be on the semiconductor industry? First, we see a dramatic increase in market share for this type of circuit. Secondly, we expect the merchant share of the ASIC market to increase dramatically. Thirdly, we expect some of the new ASIC suppliers to enter the top rank of the semiconductor industry. And finally, we expect that distributors may lose market share as the ASIC market develops.

This pie chart shows the forecast split of ASIC and standard integrated circuits. Notice that in 1980 the ASICs comprised about one-fourth of the semiconductor market, but it is forecast that they will reach half of 1990s integrated circuit market in the U.S. This 50 percent share includes the production of both captive and merchant semiconductor manufacturers.

The capitve share of the application-specific integrated circuit market is shown on the next slide. In 1980, almost two-thirds of the ASICs in the United States were manufactured by captives like IBM and Western Electric. We are forecasting that by 1990 the situation will reverse, and two-thirds of the ASICs will be built by merchant manufacturers. We believe this is a very dramatic shift in market share and make this forecast with a little trepidation.

First of all, the merchant manufacturers are very sensitive to the growth of the ASIC market, and most of them have made some kind of strategic plan to participate in this market, even if it is only in the gate array segment. Secondly, we believe that the captive manufacturers will want to retain a manufacturing capability in order to stay abreast of the semiconductor industry, but that they will farm their heavier volumes out to the merchant semiconductor industry. Our forecast makes allowance for substantial growth of captive manufacturing. In fact, the growth in captive manufacturing is equal to the growth in the semiconductor industry as a whole - namely, 18 percent compounded. Some of the growth in merchant share will come about through captives entering the merchant market as NCR did last year.

This slide is attempting to illustrate one of the significant facts of the application-specific integrated circuit, namely, that a lot of customer-handholding is required on the part of the suppliers. We've put vendor-designed circuits at the top of the thermometer as being the ones that require the most handholding. Standard products are at the bottom, and intermediate products are arranged in between. Notice that customer-owned tooling (wafer foundry business) is much more like the vendor-designed circuit than it is like the standard product. We believe that it will be difficult for a firm with a standard product mentality to operate in this marketplace, and that as a result, the firms with a service mentality will have an opportunity to increase their market share substantially.

Distributors typically supply about a fourth of the dollar volume of semiconductor components shipped, and typically supply about 90 percent of the customer accounts. Distribution is a business that depends partly on the savings in inventory costs that result when many companies' inventories are shared by a single distributor. Clearly, this part of the distribution business is not applicable to the application-specific circuit since such a circuit can only be shipped to a single customer. However, there still will be a need for ASICs among the several hundred thousand accounts of smaller customers. This part of the market is a distributor-like segment. It's essentially up for grabs at the current time because no one has figured out how to service smaller accounts with custom or semi-custom circuits.

What will the impact of ASICs be on semiconductor users? We submit that a new vendor interface will be required, that the rate of production conversion will be more important than the cost of the components, that significant engineering training will be required, that new product strategies will be required, and that new market opportunities will be opened.

The new vendor interface is summarized on the next slide. Notice that the loyalty between vendor and customer is much changed. The probability of price-cutting is quite different, and the amount of vendor support required is quite different. Purchasing departments wil have to make significant adjustments to work with this new interface.

The integrated circuit content of systems has been increasing rapidly in the past decade and we forecast that it will continue to increase rapidly. This slide shows our estimate of integrated circuit content as a percent of sales. Many companies are in the 5 to 7 percent range today. We expect the content to increase in the future. Integrated circuit content is a measure of the degree to which VLSI has been applied to the system problem. The higher the integrated circuit content, the lower the total system cost is likely to be. Therefore, for the systems company the race is to integrate its requirement as fast as possible so that costs can be reduced. Converting designs to VLSI is much more important than getting the best price on old components. Those companies that survive in the future will have the highest rate of design conversion. It's interesting to note that our statistics indicate that IBM is one of the leaders in this regard.

Clearly, someone is going to have to design all these circuits, and we at DATAQUEST believe that the engineers will have to come from the great unwashed multitude of TTL designers that currently work for systems companies. Somewhere between 1,000 and 2,000 engineers are today qualified to design integrated circuits. This slide shows the degree to which the number of qualified designers needs to increase in the next decade. Notice that the increase in new designers is an order of magnitude and someone is going to have to train all these people.

Significant reductions in product cost by 70 percent or more lead to a need for new product strategies on the part of electronic equipment manufacturers. If the product cost is reduced, it's likely that the selling price will eventually have to be reduced also because of competitive pressure. This, in turn, will require that new marketing channels be developed, and this can be a significant effort on the part of a manufacturer. The other alternative is to add features to the product so that the cost remains relatively constant and, therefore, the selling price remains constant. However, this will require that the sales force and customer base be upgraded so that they can understand the new product features and make use of them. Either strategy is likely to be difficult.

Our forecast says that we're going to product a lot of gates to VLSI. If all of today's TTL were converted to VLSI at 10,000 gates per chip, the production would not require one wafer fabrication module. In fact, we estimate that half a module would do nicely. The implication then is that new applications will have to be developed that use a lot of semiconductors. Some of these applications are shown on the next slide, and we certainly hope the user community is working hard to use up gates. Still, these applications will open new markets, and these new markets will represent opportunities for both new an old companies.

As a means of seeing this impact I went back and got the history of the RAM business as shown on the next slide, which gives the number of bits shipped as a function of year. I can remember in the early seventies wondering if we would ever ship as many bits of semiconductor RAM as of core. This was a big issue and that milestone was achieved somewhere between 1975 and 1976. We all expected the core market to be severely impacted. And, as a matter of fact, it was. However, the number of bits of core being shipped did not decrease significantly for at lease five years after the crossover point. As RAM continued to decline in price, new markets were opened which far exceeded the core market in the number of bits required to service them. These new markets consumed far more semiconductor capacity than the core market could ever have hoped to consume. We believe something of the same sort will occur in VLSI.

The next slide shows our market forecast converted to gate count. You'll notice custom LIS has already achieved crossover and we're predicting that the gate arrays will account for more gate shipments than TTL somewhere between 1985 and 1986. The products that use those VLSI gates will enjoy significant market penetration.

One is always talking about gate-to-pin ratio. We believe that gate-to-pin ratio is a function of the way the system is designed. Standard products typically have a high gate-to-pin ratio as typified by microprocessors and RAMs. The microprocessor is the top of the standard product line and the RAM is the bottom of that line. Merchant gate arrays tend to be developed to work with architectures that favour TTL pinouts, and they tend to have somewhat lower gate-to-pin ratios than do standard products. Finally, loose or parallel designs are those that are developed with no concern at all about the gate-to-pin ratio. Most of IBM's designs fall in this region or above it. It's interesting to note that IBM has provided its designers with as many pins as they want through advanced packaging technology and that this has led to a very low gate-to-pin ratios.

We believe that pins will always be expensive and that, as time goes by, engineers will develop new architectures that migrate towards high gate-to-pin ratios. We also believe that there is no formula that gives an accurate gate-to-pin ratio within even a factor of two or three.

The next slide shows some differences between the typical customer for gate arrays and the typical customer for more sophisticated ASICs. We believe the gate array customer is a bread-and-butter designer using traditional design methodologies, whereas the structured topology customer is looking for new architectures and breakthrough products.

Microprocessors have been very successful because engineers could completely develop their systems and troubleshoot them and have them working without ever waiting for a semiconductor manufacturer to produce a new circuit. This is because development systems and in-circuit emulation were available as well as field-programmable devices. Notice that the other technologies do not have all of these capabilities available but may be developing them in the future.

What is the major bottleneck to achieving the swing to application-specific integrated circuits? The bottleneck is circuit development. In this slide, we show the revenue stream from a typical gate array design, with the total revenue being about $150,000 and being developed in years three through seven in the life of the product.

In the next slide we make some assumptions about the number of gate arrays that can be developed per year in a typical company. I've circulated the number "200" because, to my knowledge, no merchant company is yet doing more than 200 gate arrays per year.

The next slide shows the revenue that would be produced by this stream of designs. Notice that it takes nine years before the company's sales reach $100 million. So this leads to the major question: what is needed? And I submit that what is needed is Visicalc for IC designers. It has to be made very, very easy for a TTL designer to learn how to work with application-specific integrated circuits, to convert this design to a pattern generator tape, to convert this tape to a mask, and to convert the mask to wafers and then packaged units. We're not there today, and there's a good deal of work to be done. In the process of doing this work, some people will find themselves very, very successful suppliers of application-specific integrated circuits.

The last slide gives our forecast for the application-specific integrated circuit market, and we forecast only the merchant segment here. Customer-designed includes any circuit in which all mask layers are changed and in which the customer controls the design. It also includes CAD designs in which the CAD system is owned by the vendor, but operated by the customer. Vendor-designed includes the traditional integrated circuit in which the vendor takes a logic diagram from his customer and develops a VLSI chip. Digital semi-custom includes all gate arrays. Linear semi-custom includes all linear semi-custom circuits. ROMs are self-explanatory. Field programmable devices include PALs and PLAs, but do not include ROMs.

"APPROACHES TO TELECOMMUNICATIONS SYSTEMS INTEGRATION IN SILICON"

Bob Broomfield, Alan Aitken

Mitel Corporation, Canada

When integrated silicon technology is first applied to an industry segment or product area it is usually on a small scale. Once started, however, the penetration and complexity of the silicon into that area grows exponentially with time. From the introduction of transistors to the consumer radio, through military programs and digital computers, this pattern has been repeated. Each program of integration adds its own impetus to the technology drive, advancing the integration density, placing new demands on the technology and developing new techniques to best utilise the available circuit density.

When technology enters a new product area, the product market becomes more competitive, functionality increases, prices fall, demand increases, semiconductor vendors are attracted by increased component market, component competition increases and the upward technology spiral accelerates. A second spiral co-exists with the first, as the advanced technology's increased density and speed allow the use of alternative techniques previously not practical or cost effective. When the technique allows higher functionality or lower cost, there is an increased demand for it. The technology again is pushed to allow new, wider applications of the new techniques. The technique is further refined to exploit the technology and the spiral progresses. The rapidity with which switched capacitor filter functions displaced the still growing integrated active filters and arrested digital filter development (temporarily) in many application areas, illustrates the effect. Switched capacitor techniques technology and CMOS technology, in conjunction with the telecommunications industry requirements, complete the elements of a dual spiral. Some current predictions say the popularity of the switched capacitor techniques will diminish, as the technology, it has contributed to advancing, matures at 2 to 1 micron geometry levels. At this point the digital filter is likely to become the more cost effective technique. Despite the attack by switch capacitor techniques in some applications DSP has already rallied and the microprocessor like digital signal processor is already advancing along its own technological spiral.

This scenario is not new, the process is as old as technology generally, only in microelectronics is the speed of development so dramatic, and at this point in time the telecommunications industry is meshed into a multiple spiral with CMOS technology, the microprocessor, (NMOS technology), TDM techniques, DSP techniques, switched capacitors, and a host of computer-aided design tools.

The microprocessor/memory/NMOS forms one of the most dramatic technology spirals and has been a major contributer to the advancement of telecommunications systems by providing low cost processing for stored program controlled (SPC) equipment and the technology to make time division multiplexed switching systems cost effective, taking digital telecommunications systems from paper into reality.

This same combination is responsible for the dramatic growth in the use of computer and the evolution of distributed data processing, escalating the requirements for data communications. Now in the early 80's those two requirements, voice and data communications, are becoming closer intertwined as digital voice switching and transmission meet similar requirements for data. This is most apparent in the business/office equipment environment. The digital telecommunications systems to service these types of needs will be used as examples later in illustrating specific applications of silicon LSI/VLSI technology and the resulting components.

The rapidly moving interaction of design technique, technology and system requirement are, in 1983, bringing an expanding range of components to the marketplace.

-Second generation speech codecs are available having been merged with the anti-aliasing filters. These LSI mixed analog and digital devices use sampled data techniques with switched capacitor implementation to provide two complex 5th order bandpass filter functions. 8 kilo-sample per second companded A/D and D/A conversion is performed with a small signal resolution equivalent to a 12 bit linear converter. Serial 2M bits/sec digital I/O allow direct multiplexing on time division multiplexed (TDM) highways and the device frequently incorporates a precision reference or digital control function. The device typically in CMOS may dissipate only 40 milliwatts when operating and may be housed in a chip carrier less than a centimeter square.

-The data modem has also been integrated into single chip form (for lower data rates). Although something of an anacranism when discussing digital telecommunications, this device provides conversion of asynchronous digital data streams to and from a modulated voiceband carrier for transmission through an analog telephone network.

-The local area network, or LAN, has itself only recently become a bus word, yet VLSI parts are becoming available incorporating not only the hardware interfaces but also the protocol converters.

-The central office and PBX line interface is being approached in a variety of ways by merchant semiconductor companies but beyond the codec, little is yet in the marketplace.

-Back in digital telephony, the time division multiplex switch is in the marketplace as a single integration circuit capable of switching up to 256 simultaneous 64k bits per second, data or PCM voice signals, from time division multiplexed I/O. This type of device generally interfaces directly to a microprocessor and uses up to 40,000 transistors to replace 8" x 12" card of MSI/LSI of 1980 vintage with a 40 pin package. Accompanying power dissipation being reduced by a factor of between 10 and 50.

-A wide variety of MOS/CMOS signalling, and a few bipolar line interfaces for both analog and digital transmission telephone sets are now in the marketplace. As yet, except for the older first generator pulse dialler, MF generator and tone ringer, there are as many approaches as there are part numbers.

-Complete and large sections of digital telephony interfaces are starting to appear, in some cases closely resembling complex microprocessor peripherals. These provide transmission and signalling facilities for digital telephone subscriber lines, T1 or CCITT 30 channel multiplex systems, areas currently commanding a strong interest from the data world.

These are only a few of the types of devices appearing as "off the shelf" components in the marketplace. Many others, either designed for telephone or data communications are appearing as specific functions or as parts in a microprocessor peripheral family. Some specific devices will be looked at in more detail later, in relation to their use in the system.

The advances allowing these type of devices to be produced are not attributable to the fabrication technology alone. There has been dramatic advances in CAD hardware and software. Design of I.C.'s is no longer such an exclusive art, thanks to dissemination of design expertise and these CAD advances. Further, the concepts of "semi-custom" design can dramatically reduce the development costs and lead times to personalized silicon. The system designer today is not nearly so constrained in his architecture by the use of "off the shelf silicon". If he believes in a more cost effective architecture, there are many opportunities to commit it to silicon successfully, if, he understands all of the implications and is willing to commit the time and money at the development stage.

For the designer defining system architecture and the circuit designer to implement the functions within it, choosing the most cost effective approach is becoming a lengthy and costly procedure due to the variety of approaches available. The microprocessor, from the 32 bit mini-computer replacement to the 4 bit micro-controller compete with high density dedicated VLSI devices which in turn compete with even more specific high density VLSI custom circuit approaches, for places in current system designs.

This paper will not try to address the issues relating to microprocessors versus specialised components as this justifies a paper in itself, but will continue to look at the choices between "standard" "off the shelf" components and custom solutions. Primary emphasis will be placed on partitioning systems for integration in silicon. It is felt that most of the criteria relating to the use of "off the shelf" components are well understood and primary emphasis is therefore related to custom design.

Semicustom design is basically a technique to provide custom parts cheaper, faster or with a lower level of I.C. expertise than full custom. As more detailed coverage will be given to semicustom elsewhere in the program it will not be dealt with further here but considered as a subset of custom design for the purpose of the subsequent discussion.

It should be noted that although the subsequent discussion centers on digital telecommunications, the principles involved apply to many areas of the electronics industry.

When approaching the question of "off the shelf" vs. "custom components" it is worth bearing in mind that the "off the shelf" part is somebody else's custom component, or at least a function of their particular partitioning of the system for which the component is intended. Generally the degree of integration of a function and the partitioning criteria for the device boundary depends on the ideal partitioning within a given architecture; inability of the technology to handle the parameters beyond the pin interface; or economic limitations or the silicon area required at the time of the design. Depending upon the device's positioning with respect to the 'state of the art' in its technology, and development of new design techniques, its partitioning may or may not be optimium when the user considers the part.

The prime aim of integrating systems in silicon (or in any other way) is to obtain the optimum functional performance for minimum cost of the complete system. Having partitioned a system appropriately into functional blocks, realizable as components, the same criteria applies to the components. That such components are a set of sub-functions of a particular system architecture sometimes tends to limit one aspect of cost reduction. That is, the part may not match the requirements of other systems requiring similar components and lose some of the advantages of high volume production. It is common at this point for merchant component manufacturers to add additional secondary functions to the device in an attempt to address a wider market. However, for the end user with one particular application, this can dilute the cost effective use of the silicon he is buying. A specific example of where this can occur is the speech codec which will be examined later. This point is made to illustrate one prime factor in partitioning systems for integration using LSI. It is important from the start to define the complete range of systems for which the integrated circuits are intended to be used, and partition them simultaneously. Partitioning one system to generate components then modifying them to accommodate other applications, to provide economies of scale, is often counter-productive when aiming for a minimum cost system.

This concept of the component (silicon) as distinct from the system is a common cause of non-optimum LSI devices. Frequently an appropriate approach to system partitioning is to assume the complete system is integrated on silicon. It is then partitioned in accordance with fitting the circuit function to available technology with a strong emphasis towards minimising the number of interconnections crossing the partitions (i.e. pins and output drive circuitry are an expensive overhead and need to be minimised).

WHY USE EXISTING "OFF THE SHELF" COMPONENTS

1. Proven operation.
2. Cost based on economies of scale over wide customer base.
3. Short timescale to hardware being available.
4. No direct component development cost.
5. Established reliability.
6. Component may provide built in compliance with end equipment specifications reducing system development learning curve.

WHY USE CUSTOM LSI

Obviously the primary reason is cost reduction which comes from the following:

1. Reduced system hardware: smaller racks, fewer cards, etc. for a given architecture.
2. Smaller power supply requirements.
3. Reduced interconnection requirements; PC track, cable, connectors.
4. Fewer component insertions.
5. Increased reliability.
6. Lower parts inventory.

These savings are due to:

1. Improved system architecture (not limited by available component).
2. Higher integration of sub-functions reducing direct component costs.
3. Direct interfacing of functions reducing requirements for "glue" components.
4. Higher level of integration reduces the number of input/output drivers required to interface via PCB - reduces power requirements and silicon area.
5. Appropriate choice of technology for function optimizes power requirement, simplifies interfacing and reduces number of components.
6. Reduction in package and pin count, improves reliability, reduces insertion and interconnection requirements.

TECHNOLOGY

One of the first aspects of designing LSI into a system is to choose the technologies to use. Generally the wider range of technologies available then the better the individual fit of the end components technology to the functional requirement of the system. To choose the optimum silicon technology for design of a large system requires a conceptual understanding of all the competing technologies and weighing their advantages against the strengths of one's own organisation.

The ability to use a wide range of technologies is generally limited by the experience, and in certain circumstances by the manufacturing capability which is available in-house. Effectively partitioning large systems using many technologies requires a wide range of expertise (expensive), and access to a wide range of production facilities. Few companies choose to use a full range of technologies on a custom basis but choose one or two technologies which fit the systems' needs most closely, and rely on standard components and other integration techniques in areas where this is appropriate.

The three principle technologies which could satisfy the bulk of the telecommunications LSI needs are listed in Table 1. Fig 1a shows the primary capabilities of a variety of technologies.

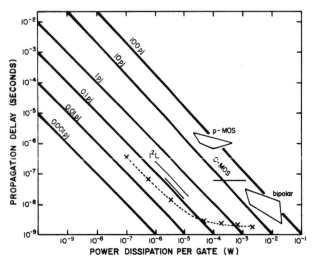

Fig 1a RELATIVE PERFORMANCE OF VARIOUS TECHNOLOGIES. THE X PLOT IS FOR ISO-CMOS with 5µm GEOMETRY.

In comparing MOS technologies, CMOS is superior to NMOS, particularly in power and tolerance to the operating environment. The superior analog characteristics of CMOS op-amps used in switched capacitor designs such as codecs is a very important factor in the preference for CMOS which is now becoming prevalent in the telecommunications industry. The use of isoplanar techniques has resulted in CMOS logic densities and gate delays similar to those of NMOS. The wide range of standard CMOS components which are available complements the custom CMOS designs which are at the heart of a modern system.

I^2L is the sole bipolar technology which, in principle, offers the potential for integrating moderate density digital logic circuitry and analog circuitry using resistor ladder techniques. While I^2L requires low power, the technology requirements for analog result in lower density digital circuitry. In addition, it is not possible to achieve high speed operation, (gate delays less than 10 nanosecs) unless one uses high speed bipolar techniques which are incompatible with the analog requirements.

The choice of which technologies are chosen is generally governed by two factors:

1. The primary requirements of the systems being designed, i.e. power requirements, operating speed, typical functional density, etc.

2. Availability of appropriate technologies.

For example, Mitel primarily manufactures telecommunications switching equipment. This requires a) mostly low - medium speed logic (less than 10MHz) b) analog/digital conversion components, filter functions in telephony bandwidth c) majority of circuitry less than 15V operation with requirements to 200V d) emphasis on low power consumption e) small system size f) high reliability g) medium - high functional density h) anticipating wider use of DSP techniques in the future there is a requirement for very high density high speed logic. To meet these needs Mitel put in-house "metal gate" and developed "high speed silicon gate" CMOS processes (5 micron production process started in 1979), the latter being the more important for meeting the evolving needs of the Company. ISO-CMOS has proven to be a very flexible technology satisfying most requirements for analog/digital telecommunication systems and continues to be evolved to meet future needs. Currently:

1) The process cannot address high voltage line interfacing requirements.

2) A policy has been made not to develop 'state of the art' memory and microprocessor devices.

The decision not to directly address (2) is based on the principle that many component companies are placing a good deal of expertise and effort with 'state of the art' technology on these very expensive and lengthy developments and it is appropriate to simply buy such components from the merchant vendors. This helps to maintain the development effort focused in areas where a high degree of unique system expertise can be applied.

High voltage bipolar processes are being investigating in relation to (1) but to date these areas are covered by the use of available MSI/discrete components which can handle the voltages required. A further degree of integration is provided by use of thick film technology as a packaging technique i.e. surface mounting using small form factor components and chip carriers. A further advantage is the provision of low cost precision resistors frequently required in the telecommunications functions where high voltages are present.

SYSTEM SPECIFICATION	CMOS	NMOS	I^2L
Line Powered	Nanowatt-Quiesscent $P = CV^2f$ - Dynamic	>10 Micro-watt/gate	<Microwatt at Low Frequency
Variable Loop Lengths (Variable Voltage)	1.5-12V	+ 10% V	1.1-10 Volts
Environment-Central Office Controlled Outside Apparatus-Extreme Temperatures	-55°C to 125°C, Digital Design Sensitive, Dialogue	0°C-70°C	As Per CMOS
Noise Immunity	30% V	+ 10% V	Typ. 0.5 Volts
Analog/Digital	Excellent CMOS op-amps	Moderate op-amps	Bipolar op-amps
Rel. Manufacturing Cost	1.2	1.0	1.2
On-Chip Memory	CMOS-Low Density CMOS-NMOS High Density	High Density	Limited Memory Capability

TABLE 1   TECHNOLOGY CAPABILITIES

By contrast a manufacturer of high speed computers may choose ECL and NMOS as the most appropriate technologies to meet his custom LSI requirements due to distinct requirements for high speed and distinct second requirements for high functional density.

It should be emphasized that although the choices of technologies is important, it is not likely to be a one time decision as the needs of the industry and the advances of technology could completely reverse any such situation in only a few years.

To summarize for a moment, the two primary steps in defining Custom LSI:

1) Define the range of systems for which the components are to be used.

2) Choose a subset of the available technologies which address the majority of these system requirements.

The latter involves taking account of:

- volume requirements (processing costs)
- analog/digital mix of functions
- operating speeds
- power requirement constraints
- voltage specifications outside of designer's control
- Physical size considerations (w.r.t. material cost and system market constraints)
- environmental conditions
- typical complexity of functions applicable to realization in one technology
- requirements for precision passive components or time constants
- current drive requirements

PARTITIONING THE SYSTEM

Having defined the system, and chosen suitable technologies, it is necessary to partition and sub-partition the system to a level appropriate to integration as single components if this is viable. When producing a function by partitioning, the system and function should be viewed in close detail paying particular attention to the interface created, circuit complexity enclosed and suitability of the enclosed circuitry for implementation in a single technology. This should take account of:

- functional complexity
- applied voltages
- interfacing requirements (drive, speed, functional standards, loading)
- total pinning requirements
- operating speeds
- component precision
- number of occurrences of the function (volume requirements)
- similar functions with minor differences in other product areas
- board level complexity (where a function is replicated a number of times on one board it may be appropriate to share elements of the functions' requirements at the board level i.e. reference or timing circuits)
- testability of function

Two points should be made which are often overlooked in initial partitioning i.e. large I.C. packages tend to be expensive; and testing can be a significant proportion of a device's cost.

The latter case applies to both production and development. Test development can be as expensive as the cost of semicustom design or a significant proportion of the cost of a full custom design.

DIGITAL TELECOMMUNICATIONS SYSTEMS

Having looked at some ground rules in partitioning for defining LSI devices, and previously stressed the necessity for defining the total score of the systems to be addressed by the components, I will be more specific about what "DIGITAL TELECOMMUNICATIONS SYSTEMS" are for the purpose of this discussion.

Traditionally telecommunications was basically involved with transferring information as voice or typed messages. As stated previously, in recent years the advances in technology, growth in the use of computers and changes in telecommunication systems have made the requirements for transferring data very similar to those for voice communications. There are many differences in the requirements of data and voice communications but these are generally outweighed by the similarities when looking at data and digitized voice communication. A digital telecommunication system can be defined as any system using digital techniques for the transmission, or switching of voice or data and its associated input and output terminal equipment, usually excluding any processing functions not associated directly with the transfer of information.

The primary elements of voice and data systems are shown in Table 2.

	VOICE	DATA
Terminal Equipment	Telephone	Terminal (modem) or Processing Interface
Transmission System	Analog/* Digital { Line Optical Radio	Analog/* Digital { Line Optical Radio
Switching System	Analog*/Digital	Analog*/Digital

Table 2   PRIMARY ELEMENTS OF A DIGITAL TELECOMMUNICATIONS SYSTEM

*When dealing with individual switching or transmission systems these would not be considered but must be considered in terms of interfacing.

It can be noted that, with the exception of the terminal equipment, the broad requirements for voice and data can be very similar, although obviously detailed signal conditioning will be considerably different.

What are considered to be currently the three most important equipment developments in digital telecommunications are the Digital Transmission Special Set (telephone), Channel Banks (and associated transmission links), and Digital Switches (PBX and CO).

Figures 1 to 3 show primary partitioning of each of these types of systems for potential LSI realization.

It becomes immediately apparent that since the PABX (or Public Exchange) has to interface with and switch information from almost all other equipment or telecommunications systems, it is the most complex system in terms of the variety of interfaces involved. Therefore it will be used as the basis for further discussion. The more common requirements are shown in Figure 3 and a more definitive list of the required interfaces is shown in Table 3.

There are many other general requirements placed on the PBX, some of the more important are listed below:

- high compatibility (low blocking probability)
- potential for high level of redundacy in common equipment
- self and remote diagnostic capability
- easy maintainability and installation
- low per circuit cost

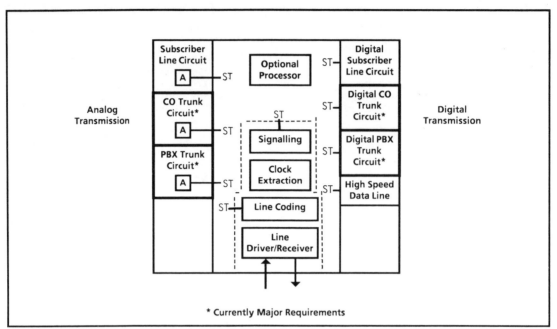

FIG. 1 - **BASIC PARTITIONING FOR FLEXIBLE CHANNEL BANK** (Multiplexed Digital Transmission Link)

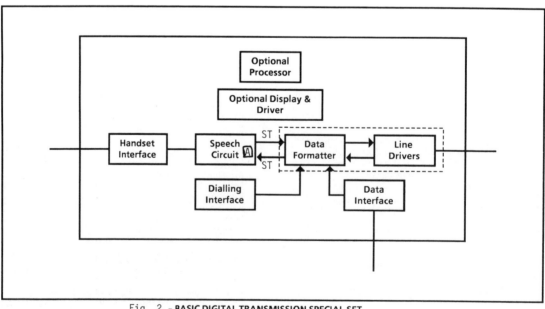

Fig. 2 - **BASIC DIGITAL TRANSMISSION SPECIAL SET**

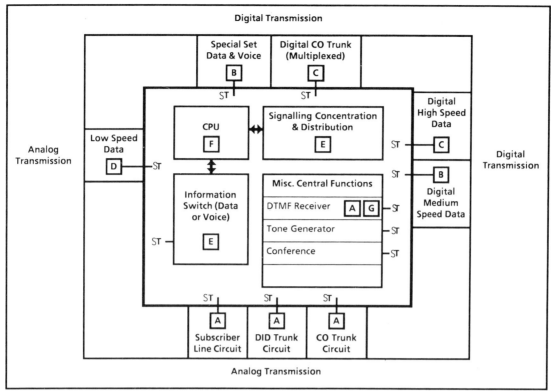

Fig 3 - BASIC PARTITIONING OF DIGITAL PABX

- low power requirements
- small physical size
- wide variety of software programmable facilities
- expandability
- direct inward dialling capability
- evolvable to accommodate new requirements and improved technology
- accommodate variety of signalling schemes
- provision of operation service

It can be seen from the above that the digital PABX is a complex machine and also that a primary partition line in the system is between the various terminal equipment, per line, interfaces and the common equipment. Provision of a unified interface between any of the individual line interface functions and the common equipment is of prime importance. Two types of information cross this interface; the information to be switched and internal control and signalling information. A major step in simplifying LSI partitioning due to standardising device interfaces can be made by adapting a common hardware protocol for both the switching information and internal signalling. Minimisation of interfacing and pin requirement can be further enhanced by adapting a serial transmission mode for this interface.

The basic interface between any per line and the common equipment would consist of four identical connection paths i.e. bidirectional information and bidirectional signalling. Further minimization of physical connection paths can be made by defining these interfaces as time shared serial highways.

Having defined such an interface any further partitioning of the central function is simplified by applying the same interface between sub-functions of the central function.

An example of this approach is Mitel's ST-BUS based architecture of the SX2000 I.C.'s system. This bus is similar in format to the CCITT 30 channel multiplex transmission format and the output format set of some first generation codecs, i.e. it is a 2.048 MHz serial data stream divided into 128 microsecond (256 bit) sequential frames. The bus operating synchronously with either a 2.048 or 4.096 master clock. A second timing waveform provides frame synchronisation. Each frame is generally (but not exclusively) considered as consisting of 32 sequential 8 bit words (32 x 64k bit channels). The ST-BUS forms the basis of the data transmission path, multiprocessor messaging system and a TDM, implied address, control bus.

As shown in Figure 3, there are four major functional devices in the central equipment; processor, switch, internal signalling concentration/distribution and common miscellaneous functions. As discussed previously, the processor system is not considered applicable for custom LSI, thus a second interface is defined by the processor bussing system. Two major LSI functions can now be identified:

1. Digital switch with direct processor compatible control interface and switching ports compatible with central/per line information protocol (ST-BUS)*.

Terminal Equipment	Tranmission Mode
Voice only Telephone	Analog/Digital
Advance Feature Telephone	Analog/Digital
Voice + Data Telephone	Analog/Digital
Data Terminal Low Speed (Equipment/ Computer)	Analog/Digital
Data Term. Medium/High Speed (Equipment/ Computer)	Digital
Multiplexed Voice Trans. Link	Analog/Digital
Multiplexed Data Trans. Link	Digital
Central Office (Various)	Analog/Digital
Other PBX	Analog/Digital

Table 3 EQUIPMENT TO WHICH DIGITAL PABX INTERFACES

2. Data concentration/distribution function with direct processor compatible data interface, with signalling ports compatible with the central/per line signalling protocol (ST-BUS)*.

The two interfaces required on both devices are essentially the same. Further, the functions they perform can be very similar. As a result a single LSI device can be defined to provide digital switching and signalling communications. The device interfaces directly with the processor used in the CPU and all other interfaces via the ST-BUS. This takes care of the major central control functions appropriate to LSI. The miscellaneous functions can be treated individually, similar to per line functions. The CPU utilizes standard functions, assuming all appropriate functions are available to meet the system architecture requirements.

The per line functions are somewhat self-partitioning as major functions. A finer degree of detail is necessary to define how to partition within the major function to define specific LSI devices. To look at each function in detail is not practical within the constraints of this paper so only one family of per line functions will be addressed in detail.

As a major criterion when judging custom LSI feasibility is volume i.e. economies of scale are very important in producing a custom part at the required cost. The current highest volume interface will be used for further discussion.

THE ANALOG LINE INTERFACE

Current digital PABX's operate in a predominantly analog environment i.e. utilizing standard analog transmission telephones and working to analog CO's. As a result, the most common interface is the analog speech circuit. The circuitry associated with these interfaces comprise the highest volume single function in the system. A large proportion of this function also exists for the digital transmission telephone but the function is transferred from the PABX to the telephone set.

When examining the various types of line interface, subscriber, trunk, DID trunk, etc. one specific function is noticeably common (identified as 'A' in Fig. 3). This is the Speech Codec and its associated filtering functions. As such, this component has been the first to be seriously addressed by the merchant semiconductor manufacturer.

*ST-BUS will be used to refer to the defined serial information and signalling interface for the sake of brevity.
ST-BUS is a Mitel Corporation trademark.

FILTER/CODEC

With regards to earlier comments in defining how the line circuitry and filter/codec functions should be partitioned for integration, all systems likely to utilize them ought to be taken into account.

Line circuits and/or filter codecs occur commonly in the following:

-PABX
- Analog subscriber circuit
- Analog trunk circuit
- DTMF receiver cards (not exclusively)
-Channel Bank (for PCM Multiplex transmission link)
-Digital Transmission Telephone or Special Set

In each of these functions the most appropriate technology for the codec/filter function is silicon gate CMOS. The partitioning of the codec/filter function within the line circuit is defined on one side by the line circuit physical interface to the backplane. In the opposite direction integrating towards the line from the filter becomes limited by the technologies ability to handle high voltages. In incorporating other functions the partition line is initially not so obvious.

If the speech circuits for each of these systems are reviewed in detail their requirements vary significantly, particularly in relation to the analog circuitry and signalling requirements. The essentials of the filtering and codec requirements are similar but there are still some significant differences. Some major considerations are illustrated in Table 3.

It can be seen that three different systems require similar types of codecs but there are some significant differences when considering LSI design. Two obvious alternatives exist: three different codecs could be produced to meet individual needs; or alternatively one device could be designed to meet all requirements. The former is expensive in development cost and product support, the latter inefficient in the use of silicon for any particular system.

The following are the major points:

- PBX and channel banks benefit from a reference shared on a per card basis, the digital set by an on chip reference.

FUNCTIONAL CONSIDERATION	PABX	CHANNEL BANK	DIGITAL TRANSMISSION
Per Card/System	Many/One Per Line	Many/24 or 30	One/One
Transmission Spec	Standard	Standard	Standard/Possible Flexibility
Signalling	Line Cct. to Internal System	Line Cct. to standard carrier protocol	Phone Cct. to proprietary line protocol
Clock Source Tx/Rx	System	System/Line Recovered	System/Line Recovered
Tx Rx Synchronous	Synchronous	Asynchronous	Synchronous to Asynchronous
Input/Output Protocol of Line Function	PBX System Protocol	Convert to standard line protocol	Convert to proprietary Line Protocol
Associated Circuit Functions -Test Function -Tx Rx Gain Control -Signalling Interface -Timeslot Assignment	Useful Useful Useful (Not TI) Could be Useful	Useful Useful Useful TI Could be Useful	Useful Useful System Dependent Not Useful
Reference	Economic Share on Card	Economic Share on Card	Economic on Chip

Table 3 - FILTER/CODEC ASSOCIATED REQUIREMENTS

- PBX and digital set do not necessarily require asynchronous Tx and Rx clocking as does the channel bank.
- all systems have similar transmission requirements.
- all systems could benefit from some additional circuitry associated with gain control, signalling (various) or on chip test functions.

To this point, two mutually exclusive circuit requirement synchronous/asynchronous transmit and receive operation and on or off chip reference, affect how to define the filter/codec in attempting a silicon efficient LSI approach.

If we now look back to the complete systems and examine where these codecs reside, the minimum digital circuit complexity area is the PBX (or public exchange) line circuit. This is also currently the highest volume requirement. Both the channel bank and the digital set (usually providing services additional to speech) have a significant additional logic requirement accompanying the speech circuits. Further, these additional functions will be present in a similar form on the associated interface circuit to a digital PBX.

To propose LSI development covering the channel bank and the digital set is a probable solution. If the points of difference between the filter codec requirements are then shifted across the partition line in the digital set and channel bank then a single codec can address all these requirements. The reference and asynchronous transmit and receive circuitry being associated with the other appropriate line circuit functions. These functions can then adopt the ST-BUS interface allowing use directly in the PBX interface for connection to the central switch, or to the codec in the stand-alone system.

The codec when defined under these circumstances can make use of a number of techniques which allows minimization of silicon area required and can still be used in each of the three types of systems without creating the necessity for discrete or MSI interfacing components.

MT8960

The MT8960 codec family was designed on this basis to give what is currently the smallest die in the industry for this function (20000sqmils). An outline of the resulting specification is given below:

- transmission specification meets CCITT G711/G712 ATT specs.
- synchronous Tx Rx digital I/O (to ST-BUS format, Fig. 4).
- digital control port (operates on ST-BUS format).
- digitally controlled drive outputs.
- digitally controlled Tx and Rx gain adjustment.
- digitally controlled test functions - analog/digital loopback.
- off chip reference.
- ISO-CMOS technology - low power operation, switched capacitor filters, switched capacitor multiplying DAC Codec +/- 5V power supply requirements.
- small die size.
- mask programmable Alaw and ulaw options.

Consistent with further integration of the line circuits, codec variants MT8960/61/62/63/64/65 (accommodating variations in companding laws and output code formats) are packaged in chip carriers for mounting on thick film hybrid circuits, as well as in standard dual in line packages. Consistent with the serial interface approach at partition lines, the PCM inputs and outputs of the codec and also the control port (DX, DR, DC, Fig. 4 & 5) operate to the ST-BUS format.

Other interfaces referred to in the previous discussion are the Digital Set Interface (identified as 'B' in Fig. 3) and Digital Trunk Interface (identifed as 'C' in Fig. 3). Their functional diagrams are are shown in Figs. 6 and 7 respectively. Again, note serial system ports are ST-BUS. The Digital Line Circuit has microprocessor direct parallel bus control, the Trunk Interface ST-BUS control. Both functions are (as shown) implemented using thick film hybrid technology, a major part of the functionality being provided by a single CMOS integrated circuit. Hybridisation providing precision resistors and packing density advantages. The high voltage circuitry is provided using discrete components on the hybrid substrate.

The previous discussion and examples attempted to stress that attention should be paid maximizing the amount of system silicon and minimize total silicon in the system. This is preferable to approaching individual functions independently when contemplating integration. Further, standardisation of interfaces at partitions is important and significant savings can be made by making this interface serial where operating speeds allow it. The former prevents dilution of the functional density of the circuit board which happens when MSI is used to interface between LSI/VLSI devices and the latter reduces the large amount of silicon area which can be wasted on large output drivers, driving PCB track capacitance.

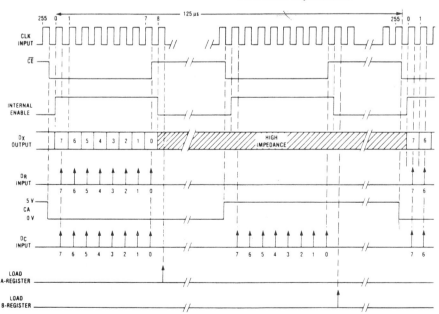

Fig. 4   MT8960 ST-BUS TIMING DIAGRAM

The principles discussed are not unique and neither is the partitioning shown in Fig. 3. However, there are a number of approaches to implementing the partitioning shown. Many semiconductor manufacturers have synchronous and asynchronous codecs available, and various forms of digital set interface. There are also various implementations of the digital switch and subfunctions of the digital trunk interface on the market. However, there are as many ways to build systems as there are system engineers and even items like speech codecs can prove cost effective to custom design if the designer's architecture warrants it. The device described above was just that situation.

As a postscript some reference should be made to where DSP technqiues seem to be applied in telecommunications. Briefly the general purpose DSP devices seem to be finding homes in smaller quantity functions such as the tone receivers, speech recognition functions, echo cancellation (satellite communications), etc. Specialised functions are being addressed to the line circuits, but in this case primarily the digital, transmission, line circuit.

Whether the choice is for "off the shelf" or custom, there is no shortage of approaches to using silicon in a system or systems on silicon.

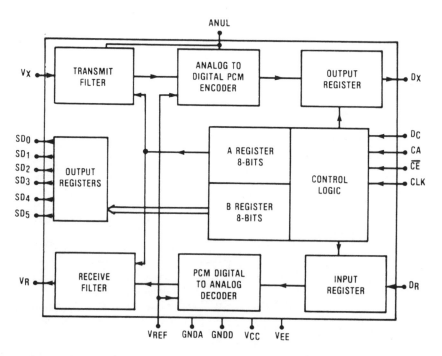

Fig. 5 MT8960 BLOCK DIAGRAM

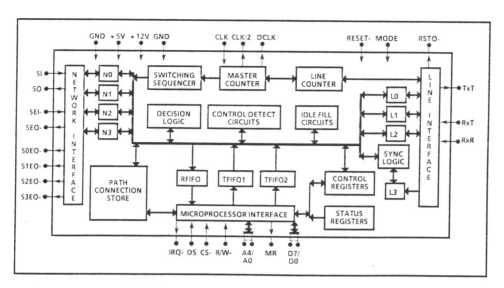

Fig. 6 DIGITAL TRANSMISSION LINE INTERFACE

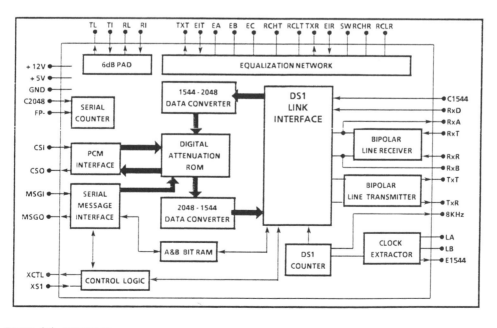

Fig. 7 DIGITAL TRUNK (1) INTERFACE

PERSONAL, LOW COST, COMPUTER SYSTEMS

J B Tansley

Acorn Computers Limited, Cambridge, UK.

INTRODUCTION

The notion of a low cost personal computer system means different things depending upon your point of view. For the purposes of this paper I propose to consider as low cost any minimal, yet usable, computer system that can be bought for less than $1,000. Clearly this is a quite arbitrary definition and says nothing about the intention of designers of more expensive systems nor of the desires of potential purchasers of personal computer systems. As a simple model it possibly comes nearer to the motivation of the general public as buyers of personal computer systems since, at the moment, most people tend to be governed primarily by the basic cost of a system rather than by its utility. For 'personal' I shall generally mean a computer system bought for exclusive use by one person at a time with the intention that the system will have all the necessary processing and storage facilities.

It is generally believed that the principle factor in bringing about the advent of the low cost personal computer has been the development of integrated circuit technology. Since the early 1970's we have moved from single silicon dies containing a few tens of transistors to those of today which can contain some half a million devices. The major element in cost reduction arises from the fact that an increasing amount of a system circuit can be integrated onto a single chip while at the same time the cost of reproducing these chips has changed to a far less degree. There is little evidence to challenge this view; the reasons why it has come about are debatable; the consequences of its existence are far more interesting.

LOW COST SYSTEMS

It was a few years before micro-processors were exploited to build cheap computer systems. The major effort in the design and manufacture of cheap personal computers such as the Apple II, Commodore Pet and Sinclair Spectrum has been dominated by the provision of the constituent hardware. It was a narrow and highly focused activity, geared to exploiting a series of basic micro-processor components such as the Z80, Intel 8080 and Motorola 6800 and their immediate derivatives. Superficially, the systems design task seems easy. Indeed there are now books that tell you how to do this stage by stage. It is largely a connectivity task and the conventional way for a commercial system builder to go about it is to design a printed circuit board and package a number of sub-system components together in a suitably attractive case. There are several reasons why this systems design task has proved to be much more difficult and complex than might be expected from an initial examination.

First, our understanding of software and our desire to entertain computational activities which may give us a competitive edge was well beyond the simple capabilities of the 8-bit micro-processors then available and still used in most low cost computer systems. Consequently, clever hardware was needed to support quite minimal low level systems and application software. This precluded a naive approach to systems design; some care and consideration was required in deciding how best to connect things together so that a system could develop flexibly without having to start again. This sophisticated approach to systems design has allowed manufacturers to develop some individuality in the kind of products that they have offered and, in the long term, may determine who is to survive.

Second, partly arising from the first problem and partly from the idiosyncrasies of the sub-system manufacturers and silicon designers there was, and still is, often as not, a functional and electrical mismatch between the various sub-system components. Designs that cope with these difficulties tend to have a significant increase in the number of "glue" components, which leads to a subsequent rise in the overall system connectivity and hence cost.

The currently favoured solution to these problems is to use a gate array technology to provide a single silicon component that contains most of the extra connectivity required to make the better system. The low replication costs of silicon products then reduces the overall system costs.

DESIGN AND MANUFACTURING ENVIRONMENT

At present there are three groups of people involved in the design and manufacture of low cost computer systems. First there are the innovative companies with systems designers exploiting low compute power micros in clever hardware systems. These companies have a low level of capital investment. The risks are quite high since the success of products depends upon a fickle public. Already the demands for better software i.e. usability, are growing.

Second there are the major sub-system designers and manufacturers. This group make things such as power supplies, keyboards and disks. The level of capital investment may be higher but the risks are somewhat lower. The main complication arises from the need to produce products to satisfy evolving technical interfaces.

Finally there are the silicon manufacturers. The capital investment in this area is very high and because there is the obvious necessity to sell in high volume the risks are also very high. A quality fabrication line can cost in excess of 20 million dollars while the design of a single moderately complex custom chip using existing methods is quoted as being in excess of 5 million dollars. Given the right 'jelly bean', costs can be quickly recovered, the wrong product can be a crippling mistake.

This design and manufacturing triumvirate would appear to be a profitable and acceptable arrangement. At the moment competition exists mainly from within each of these groupings and certainly the systems manufacturers would like to have a period of time in which to exploit their success with the 8-bit micro products. Unfortunately, or fortunately if you are a potential customer, the world of information dissemination and processing is fragile. Product stability is illusory and a variable window can change at any time. Detailed prediction is impossible but it is an interesting and worth while activity to attempt to understand this era of change. There are many ways in which this can be approached, I propose to look at two with respect to the potential progress of low cost computer systems. The first concerns the consequences of vertical integration, of one company doing the whole thing from systems to fabrication and the second concerns the consequences of developments in software. Sometimes this view is described as the balance between technological pull and market push.

VERTICAL INTEGRATION - THE RISE OF MARKETPOWER

If we relax the earlier definition of low cost as applied to personal computer systems then it is reasonable to claim that this is the area of major growth in computer systems. The cost of comprehensive computer systems is now below a threshold whereby they are now bought for the exclusive use by one, or at the most, two people. Additionally, in many companies and organizations the decision to buy such a system is at the lowest level of purchasing responsibility. The use of personal computers will soon be the way in which most information processing is carried out.

The prospect of the personal computer has stood on the horizon for a number of years. For example, the DEC LINC 8 and the work at Xerox Parc with the Alto has demonstrated quite clearly the acceptability of this form of computing. At the same time technological advancement has also led to ever cheaper systems, for example the DEC 11 series, yet these systems have been a mere replication of the batch and timesharing systems of the 60's. The possibility of a simple progression toward personal computers seemed obvious. It has taken the viability of the very low cost, low capability system to create the era of the personal computer.

It is only recently that the computer manufacturing establishment has viewed this change with any seriousness and only extends the view of personal computing from that already discussed. Fortunately there are many people who feel they have a need for a personal computer hence this market sector is expanding rapidly. Furthermore, there is little evidence to suggest that it will saturate for some time yet. The billion dollar nature of this market means that a large number of organizations would like a large share of it. In particular, the more traditional computer manufacturers would like to have a share if only because the personal computer is impacting on their present markets.

All this suggests that the market is becoming very competitive.

The competitive response during this episode of market expansion is for vertical integration of manufacturing and development. Three of the major reasons for doing this are, first, to have much better control over the whole range of production, from systems integration to silicon fabrication; second, to be in a better position to trade various design alternatives so that products can appear more attractive yet cost a minimum; thirdly, the integrated company can exercise a high degree of market power.

In summary, the low cost personal computer has created the "personal image" of computing. Its exploitation and market pressure dictates vertical integration to do this effectively.

SOFTWARE - THE TECHNICOLOGICAL PUSH

Software is something that lots of people use and very few people understand. It is also singularly the most important entity as far as computer systems are concerned. Software is at the same time the technological pull for the new and interesting things to do with computers while also being the technological push to re-organize and restructure hardware. Software appears to have an abstract and undefinable character yet we need to assess its influence on hardware. We need a view of software that shows where it stops and hardware begins.

Essentially software may be seen as comprising of programs written in a machine language L(n), which are either interpreted by an interpreter running on a (virtual) machine M(n) or are translated to the machine language of a lower machine.

Diagrammatically:-

At the bottom level programs in L(1) are directly executed by a fixed model of computation which is embedded in electronic circuitry. At this lowest level the model of computation is synonymous with the physically invariant computing machine.

With this perspective it is fairly easy to classify micro-programming, firmware, different programming languages and the like.

Within this framework what can be said about low cost computer systems based upon 8-bit processor products?

The model of computation offered by 8-bit processors is restrictive in two major dimensions. First, these machines only offer a very limited address space with respect to program code and data. The best that is available now is 64K storage locations.

Second, the instruction set of most 8-bit micros is not conducive to the efficient translation from higher level languages.

Further, these processors tend to be not very powerful. The most effective use is by interpretation or by programming directly with the machine instructions.

When 8-bit processors became available in volume, software had advanced to a computational model well beyond that offered by these devices. The fact that many of the low-cost systems do so much is a tribute to some imaginative hardware systems design that reduces their limitations to a minimum.

The situation with the new second generation of 16 and 32 bit single chip processor products is somewhat different. First, the major address space limitations have been removed and second, the basic computation model offered by many of these machines is more "advanced" and able to meet some of the needs of higher levels of computation. This should mean that the systems hardware utilizing the 16 and 32 bit processors will be easier to construct. Highly sophisticated individual hardware design will not necessarily buy you that much. It is much more important to design systems that meet the intended computational models of the micro-processor chip sets. Deviation from the accepted line could cause severe performance penalties. On the other hand some cleverness may be required to fix up the mistakes in the original processor designs.

Once these early problems are eliminated by further re-design of these 16-bit processor chips it should be very easy and cheap to construct very powerful personal computer systems. These systems will be mainly differentiated by the quality of software that they run and by the range of peripheral systems and devices accessible from these personal computers. This development poses some interesting marketing decisions. Should hardware be given away in order to create a large customer base? How effectively can the software for this market be protected?

In summary, with 16 and 32 bit processor based personal computer systems the competitive edge will change from hardware to software.

INNOVATION

I have suggested that the development and availability of software is the key to the success of future personal computer systems. Within this context what scope is there for innovation and enterprise?

Literally, innovation is synonymous with change, but within the world of information technology it has tended to mean more than just this. In particular, the notion of innovation has acquired the sense of change of at least an order or magnitude. Two types of change come to mind.

The first is concerned with the creation of new products or services that establish new and large markets.

The second is concerned with the use of new design and manufacturing techniques that result in better and much cheaper existing products.

Ideally the two mechanisms of inducing change should be combined, i.e. use new design and manufacturing techniques to produce new products. This is an unusual occurrence as a large capital investment is often required and those with the money have no notion of the risks involved.

Almost by definition, innovation is something most people have difficulty coping with. In part this may be due to the fact that they do not recognise the magnitude and scale of change implied by a new product, novel design method or manufacturing idea. Its acceptance could mean a total abandonment of previous beliefs and radical reorganization of ideas. The human ability to cope with change in a familiar environment is limited. There are a number of psychological theories that discuss this phenomenon and I will just note that it is collective human awareness that limits progress in information technology and not a lack of innovative ideas.

Also, by definition, the computing world differs from other "normal" engineering because it does not have the same kind of physical constraints and so is easily able to redefine and restructure its concepts, dismissing all that has gone before. Thus the computing and digital systems world is an abstract one. Only at the lowest levels of implementation does it have any physical realization and hence only here is it constrained by the laws of physics. This means that the partitioning and heirarchy associated with computing systems is quite arbitary and pure invention that happens to be convenient for present methods of designing and building systems.

The language-machine model discussed earlier can be used to show some examples of this arbitariness.

1. The language ADA for the Intel IAPX432 machine.

2. The language 'C' for the Berkeley RISC machine.

3. The language OCCAM for the INMOS transputer.

4. The language FORTRAN for the IBM 709.

For each of these schemes it is intended that our use of 'silicon' is at an abstract level. At present we have no way of deciding what is the best route to follow.

In summary: the level or representation of digital systems (personal computers) is unbounded above and potentially bounded below by a stable digital interface to silicon. The scope for innovation is limited only by our ability to devise appropriate models of computation. Open access to stable silicon could provide the incentive for the development of the most amazing things yet to be encountered in the whole world of engineering.

INTEGRATED CIRCUITS IN CONSUMER PRODUCTS

J. D. Leggett

General Instrument, Microelectronics, USA

ABSTRACT

Consumer products will grow only 2% this year, but will contain 12% more electronics value. Certain segments will see very rapid growth and the industry must respond to increasingly difficult demands on design resources and response time. Product designers must carefully chose electronics implementation to match their own strengths and select silicon suppliers based on their own market needs. The most important lessons for designers and silicon suppliers are the necessity for careful specifications and for realism of mutual expectations. Finally silicon implmentation must be supported by timely production.

CONSUMER ELECTRONICS MARKET

The worldwide consumer equipment market will see less than 2% growth in end product sales while the market in integrated circuits is expected to achieve almost 25% this year.

Sales of special purpose consumer chips such as for watches, games and cameras, will grow at about 12%. (See Table 1). The most straight forward inference is that the IC content of consumer products is increasing. This is supported by a 1982 study of the total IC content of various segments. Selected segments are shown in Table 2. Even segments showing slow growth, such as TV receivers enjoy rather high (18%) growth in IC content, due to pressures for added features and reduced cost. Integrated circuit consumption for more rapidly growing end markets such as musical instruments and home computers increase typically about twice as fast as the end product. On the other hand, a technologically stable product such as video game consoles, has shrinking integrated circuit content (as measured by selling price). Many of these markets are volatile and difficult to forecast. Estimates of the home computer market, for example, vary by more than a factor of two.

Telephone handsets provide a good illustration of the consumer market. The annual volume has grown from zero in 1978 to 5 million last year and is expected to reach 15 million in 1983 (2). But to compete in this market, both producers and IC vendors must respond very quickly. This year the bulk of handset sales will be low end units, and competition will be dominated by price. Next year the battle will be fought over features requiring a new generation of chips in very high volume production.

Table 1 - Worldwide Consumer Product and Integrated Circuit Sales
Source: Electronics

Market	Sales, $M 1982	Sales, $M 1983	Growth
Consumer Products	48188	48966	1.6%
Integrated Circuits	9833	12243	24.5%
Special Purpose Consumer IC's	967	1079	11.6%

A case history illustrates the volatility of the consumer market as seen by an IC vendor. Figure 1 shows a five year experience with a selection of chips for video games. Such rapid market saturation is now a fact of life in some consumer markets. So is the requirement to achieve the rapid production increases necessary to participate. The ability to achieve rapid market entry by fast design capability with high confidence of success is crucial.

Thus, while the consumer market as a whole is growing slowly, certain segments are dynamic. This presents special challenges to producers and places difficult demands on integrated circuit vendors.

Table 2 - Selected Worldwide Market Segments, 1982-1987 Growth

Segment	End Equipment Sales CGR%	IC Consumption CGR %
Television Receivers	2.0	18.2
Telephone Handset	15.0	37.4
Musical Instrument	8.5	18.0
Home Computer	24.6	43.1
Game Consoles	4.8	4.6

Sources: Electronics, General Instrument

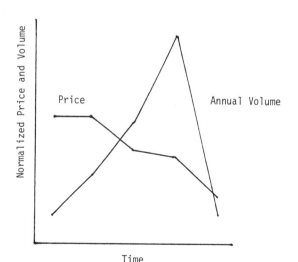

Figure 1-A Case History: Price and volume of selected chips.

APPROACHES TO MICROELECTRONICS

The consumer market is characterized by a very wide variety of approaches to electronics implementation. The silicon content is typically less than 10% of the end product sales, and often consists of no more than a handful of chips. A popular low end personal computer, for example, contains three chips.

Table 3 shows an estimate of the integrated circuit segmentation within the consumer market. Over 60% of this silicon market is microprocessor/microcomputer plus memory, with about 25% being custom and I/O devices. This division reflects the relative advantages of various implementation approaches. The relative advantages and disadvantages have been well publicized, especially by the proponents (vendors) of each, so it suffices to summarize. Software implementation in a micro gives fast implementation with ease of changes and enhancements. This approach usually requires "glue" or special I/O capability, often realized in a ULA (otherwise known as gate array) which can be relatively rapidly implementated with high confidence. Full custom gives (often) a single chip solution at lowest silicon cost at the expense of longer turnaround time and increased risk. Finally, standard cell can give much or all of the benefit of custom in less time at only slightly higher part cost.

I believe that the key lessons in successful implementation are independent of the design approach, and in todays environment of cost and time sensitivity the lessons are increasingly important.

KEY LESSONS

The key lessons in applying microelectronics to consumer products (and many others) are careful specification, and realism of mutual expectations. In addition, implementation is only successful if supported by timely production.

Successful implementation means silicon which functions properly. In order to succeed it is essential to know what is required and to be able to specify it to the vendor in precise terms. ULA implementation probably best illustrates this problem. Many ULA vendors now impose discipline on the customer, who must use the vendors tools, perhaps at a local design center. For those customers who know exactly what they require this can be very successful. But being taught the tools is one thing, and understanding the circuit and how to realistically simulate it is another. The customer must be realistic in assessing his own capabilities, and he must not let the glamour of the tools take precedence over thinking and understanding.

ULA implementations usually pass the test program the first time, but estimates for probability of first time success in the product range between 1/2 and 2/3. This is because the vendor (usually) supplies exactly what the customer said he wanted, not what he needed. This is somewhat less of a problem for custom or standard cell, because customer-vendor coupling is usually much tighter. Of course both must be prepared to make this investment. When this investment is made for ULA designs, the success rate is very high.

Testing can become a major issue because the customer may not appreciate the problem or it's implications. And specifying complete tests can be so difficult that neither wants to undertake the task. The result is very often several percent incoming failure, or even a non functioning design. Successful implementation depends on a very clear mutual understanding of testing responsibilities.

Realism of expectations is most important in scheduling. Experience shows that between 1/2 and 1/3 of ULA designs don't work correctly the first time, and it is important to schedule in a design redo, but the customer often wishes that every design step will be best case. The vendor encourages this, because he wants the business. Wishful thinking is the surest path to silicon disappointment and frustration.

Finally, the case history above is convincing evidence that production timeliness is the key to success in some consumer markets. Unfortunately, those vendors with the most suitable and fastest design capabilities do not always have the required production facilities for sufficiently fast production ramp up. Some producers use design houses and a silicon foundry to avoid this problem. Others now do their own design and use foundries with substantial production resources.

REFERENCES

1. World Markets, 1983, Electronics, 56, 125-156.

2. Push Button Boom, Barron's, June 6, 1983

Table 3 - Consumer IC Market Segmentation
(Note: includes personal computer)

	Micro	Memory	I/O	Custom	Other
1982	27%	36	8	16	13
1983	22%	40	9	15	14

FUTURE REQUIREMENTS FOR VLSI IN THE MILITARY FIELD

K.W. Gray and B.J. Darby

Royal Signals and Radar Establishment, Malvern, UK

THE MILITARY NEED

It has always been that the operational effectiveness of a military weapons system has largely depended on the performance of a human operator. Crucial actions requiring a high rate of threat evaluation and high levels of skill have often been demanded simultaneously with the operator experiencing extreme physical discomfort and fatigue. Electronic systems are continuing to evolve that greatly extend the range of human performance and workrates. Equipments incorporating signal and data processing equipments now play vital roles in such areas as target acquisition and engagement; command, control and communication and electronic warfare. Detailed examination of the hardware and software needs in systems for the 1995 and beyond timescale have been widely examined considering evolving threats and they consistently require systems with extended processor performance. There is no forseeable respite. Signal processing is thus a crucial force-multiplier, a strategically important technology.

To illustrate the military need for high performance equipments we now briefly consider three examples.

1. Multimode Airborne Radar

Military aircraft demand increased performance in their weapons systems capability to achieve their objectives. Hence, one of the most important contributions promised by IC technology development is the accomplishment of a multimode tactical airborne radar offering simultaneous air-air and air-ground capability.

For maximum effectiveness air-air modes include target detection, acquisition and tracking. Air-ground modes may include real beam, doppler beam sharpening and synthetic aperture spotlight map for area mapping, stationary target detection and weapon delivery. In addition air-ground modes for detecting moving targets and for weapon delivery may be required. Also automatic IFF, terrain following, terrain avoidance and precision navigation capabilities are suggested to reduce the pilots workload.

This multimode capability is sought through programmable processors. The radar signal processor must be able in some modes to simultaneously cover several of the following options: clutter tracking and cancellation, adaptive sidelobe cancellation for nulling jammers, doppler processing and digital pulse compression, constant false alarm rate detection techniques, etc. Each of these functions presents a high processing load which demands careful attention to algorithm-architecture mapping in order to achieve the necessary speeds and efficiencies.

In all, some twenty modes could be envisaged for future advanced tactical fighters. Notwithstanding increased reliability, fault tolerance, reduced life cycle costs and increased availability are required.

TABLE 1 - Processing requirements for an advanced multimode tactical airborne radar

	Basic Need	Current Technology	1 μm Technology
Throughput (MOPS)	300-500	30	<300
Memory (Mbit)	12	12	12
IC Count	-	1000	500
Power Diss (kW)	1	1.5	0.5
Size (cu ft)	1	1	< 1

2. Tactical Radio Transceiver

In order to provide secure, survivable, tactical communications in the forward area for the mid 1990s and beyond, radio transceivers, which are considerably more flexible than those in use today, will have to be developed. In particular the electronic warfare features must be capable of enhancement during the lifetime of the radios.

Programmable digital processing applied to tactical radio equipment design is expected to confer such requirements as the ability to alter carrier frequency, modulation type, error control coding, network management, ECCM etc. A capacity for incorporating more sophisticated processing is seen as key to extending the equipment life-cycle whilst meeting the evolving threat.

The hardware envisaged could be a single transceiver unit covering a very wide operating frequency range and suitable for both manpack and vehicle use. The processor throughput requirement is approximately 20 MOPS with up to 50 MOPS being necessary for a flexible unit. The small volume and weight required together with a need to keep the power dissipation < 1 W puts this development firmly in the VLSI category. For reference, current LSI technology would result in at least 30 W power dissipation which is clearly unacceptable for manpack use.

3. Autonomous Seekers

Current techniques for missile and munitions guidance impose severe operational penalties on the weapons platform. Beamriding, command to line of sight and laser designated systems require accurate target tracking during the flight time of the weapon. Continuous, accurate target illumination leaves the attacking aircraft vulnerable and limits the rate of target engagement. Provision of multiple target tracking makes these types of guidance control option expensive. The next improvement is lock-before-launch systems but these have limitations too.

Autonomous target acquisition and homing for future missiles and submunitions are very attractive. To achieve high performance significant advances are required in signal processing hardware and software to exploit fully the information presented by advanced sensors. The missile (or submunition) is launched into the vicinity of a target or group of targets and once within range of autonomous homing a target is acquired, identified and engaged. Ground based stationary targets embedded in natural clutter and camouflage and electronic warfare highlight the need to press developments of the most sophisticated signal processing algorithms for feature extraction for either radar or IR sensors. Data throughput and arithmetic rates will be very high and VLSI circuits are needed to meet performance, size and cost goals. The estimated arithmetic rate is 100 MOPS. The terminally guided submunition and smart shell application presents the greatest challenge for hardware miniaturisation.

TABLE 2 - Arithmetic rates required for a selection of advanced systems compared with current service systems

	Today's System in Service (MOPS)	Advanced System (MOPS)
Airborne Radar	30	300-500
Ground Radar	200	10000
Autonomous Seekers	2	100
Radio Transceiver	Analogue ≡ 10	20-50
Image Processor for Battlefield Surveillance	Non real-time	200-500

For the selections of military systems given in table 2, the significant factor is the demand of between one and two orders of magnitude increase in processor arithmetic rates. This increased performance directly determines the level of technology required since the size, weight and power dissipation constraints of the conventional unit will still apply. A convenient figure of merit for indicating the required technology level is the functional throughput rate (FTR) defined as the product of the number of gates per chip (or per Watt) and the chip clock rate. Usually for signal processing the FTR requirement can be related to the necessary arithmetic rate when given the algorithms to be executed and the machine architecture. Figure 1 relates the required average FTR to the number of integrated circuits in the processor for three of the examples previously given. Contrasted with the advanced equipments which require technology

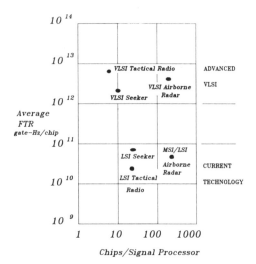

Figure 1 Relation between FTR to number of processor IC's

in excess of 10^{12} gate-Hz/chip are examples of what currently available technology, which has an FTR less than 10^{11}, can achieve. These examples are for an LSI seeker, an airborne intercept processor and for a straightforward digital replication of the analogue functions in a tactical radio processor with LSI DSP microprocessors (1).

On this figure the penalty of extra gate complexity for achieving a given performance with programmable processors has been included. A factor of fifty in FTR has been estimated (by counting processor clock cycles) as the ratio between the extremes of implementation on software processors versus dedicated processors. In practice, many programmable processors will have embedded dedicated functional blocks, eg a complex multiplier macrocell, which will greatly reduce the ratio. Hence, physically small military systems will have some degree of programmability.

MILITARY VS CIVILIAN VLSI

The growth in complexity of military silicon integrated circuits lags commercial products by several years, a reversal of the situation 10-15 years ago. Commercial IC suppliers have continuous new business incentives (timescale and cost) and are faced with high levels of competition. These factors force improvements in both IC technology and CAD and CAT systems which have resulted in high levels of silicon integration. Furthermore, the high volume production associated with successful products can make the whole operation economically viable.

The most complex ICs and rapid growth are associated with general purpose computer developments and are particularly noted for memory components, chip sets in support of video games and the now wide choice of 16 bit processors. Hewlett-Packard's 32 bit CPU with approaching 500,000 transistors on a chip, requiring 1 μm feature sizes, is a recent example of the fruits of this commercial drive

There are various reasons why military component complexity has lagged behind commercial examples. Military systems tend to take a long time to develop performance levels which are acceptable to the armed services. This is a direct consequence of the high equipment complexity and the need to operate in adverse environmental conditions. Inception to in-service time scales of over 10 years are common. There is a high design and re-proving cost and time penalty associated with change. In development, redesign of critical subsystems may be carried out at a late stage in order to meet specification. Once an equipment is in full production a low production volume, often characteristic of military systems, will ensure that design changes are not cost effective even when lower unit manufacturing costs, greater reliability and improved performance are promised through introducing new technology. The long in-service life of ~ 20 years for military systems in peacetime and the need to maintain a technology supply over most of this period contrasts sharply with the short lifetime of a commercial integrated circuit 'standard'.

How are the military users going to benefit from the dominant civilian use of VLSI? For the immediate future, design costs of complex signal processing chips may be contained by maximising the use of well characterised "macrocells" and by exploiting fully architecture-algorithm regularity (2). (Highly regular architectures should present the additional benefits of improved testability, leading to full BIT, and some degree of fault tolerance.) Thus the average design cost per gate could be substantially reduced if the necessary advances in computer-aided design, description and simulation packages are realised.

One other approach may be to design a family of "programmable" VLSI components where hardware architectures match a range of commonly used algorithms. Initially, these devices may be difficult to realise requiring a careful balance of hardware and software architectures and design of I/O mechanisms for increasing system complexity and throughput by using multiple chips per processor. However, once available, very high production volumes could achieve acceptable hardware costs while development and redesign costs would depend chiefly on software costs. Hence, it is expected that, for similar reasons to the need for a range of hardware macrocells, a range of efficient software macros would be developed as is occurring now in microprocessor work.

These predicted trends, which are already visible reflect the fact that there exists a degree of commonality in the tasks performed in signal processing. Certainly, as pointed out by Roberts (1) commonality can be demonstrated at the functional level although the application will determine the algorithm parameters, the throughput rate and the physical constraints. It is thought that identification of the range of commonality obtained throughout the spectrum of military equipment is vital to ensuring that military needs are met rapidly and cost effectively with future complex VLSI components. Table 3 shows an attempt at such a categorisation. It was assembled following a consideration of:

(1) processor performance, including throughput and programmability;

(2) physical constraints, size, weight and power dissipation;

(3) economic factors, production volume and test and maintenance factors; and

(4) algorithms employed and architecture options.

The contention is that applications within the same broad category could be furnished by chips developed for a key driving application eg one which met the most demanding performance and physical constraints.

It is believed that by reducing the number of radically new designs necessary, through appropriate use of macrocell and programmable components, military equipments force fitted to these categories can more nearly keep pace with the growth in IC complexity and technology. This would occur because the production scale of these complex devices would be sufficiently high to warrant continual upgrades.

TABLE 3

	PORTABLE SYSTEMS	MODULAR SYSTEMS	ARRAY SYSTEMS
TECHNICAL PROPERTIES			
Data Throughput	50-100 MOPS	100 MOPS	300-1000 MOPS
Programmability	high	limited	very high
PHYSICAL PARAMETERS			
Power Consumption	low	limited	limited
Size	small	small	compact
PROCUREMENT COMPARISONS			
Production Scale	high	very high	low
Cost	low	low	moderate
Examples	Man portable radios and surveillance systems	Radar/IR/Sonar missile ammunition seekers	Radar signal and surveillance image processing

REFERENCES

1. Roberts, J.B.G., "The impact of VLSI on the architectures of signal processing systems". This Seminar.

2. Whitehouse, H.J., "The interplay of architecture and algorithms for real time signal processing". This Seminar.

Copyright © Controller HMSO, London 1983

SUMMARY OF PAPER ON MILITARY SYSTEMS

W.S. BARDO

MARCONI SPACE & DEFENCE SYSTEMS LTD.

The significance of VLSI to military systems

Silicon VLSI offers the military systems designer an opportunity to improve the capability of weapons and to reduce their size, thereby enabling the launch platforms to carry more weapons of greater effectiveness. The reduction in size of the weapon is achieved in two ways, firstly, a reduction in size of the electronics; secondly, a reduction in miss distance achieved by smart signal processing permits a large reduction in size of warhead.

The development of smart weapons that can find and attack their targets autonomously permits the launch platform to stand off from defences and achieve a higher sortie rate.

While concentrating on silicon technology, it is worthwhile placing this technology in perspective: other technologies are important to the military system designer. Two examples are gallium arsenide monolithic microwave integrated circuits for compact broadband microwave processing, and millimetre wave technologies that are particularly significant to the development of small but smart anti-tank weapons

What does a military systems designer look for in the silicon technology?

Speed
High packing density
Low power consumption to avoid thermal problems and to minimise the drain on battery power
Radiation hardness

These considerations have led Marconi to invest in silicon-on-sapphire CMOS as a military and commercial technology.

The changing relationship between the systems designer and component supplier

The systems designer has been used to working from block diagram down to circuit diagrams, and constructing a breadboard out of TTL components. He may progress then to MSI devices interspersed with discrete devices.

With the advent of LSI and VLSI breadboarding becomes impossible: one is likely to be solving a set of entirely inappropriate problems. Breadboarding is replaced by simulation. The systems company needs engineers who can bridge the gap between the designer producing the functional description and the highly skilled designer who has a detailed understanding of the technology. CAD tools are critically important. Workstations, in stand-alone form or netted to back-up computers, have an important role to play in providing the degree of interaction needed to make optimum use of the rather different capabilities of man and machine. The floor-plan is all important in making a good start to the design. Here the necessary tradeoffs occur in placement and routing to satisfy both the functional and technology requirements.

The military system designer is motivated more by the need to get a short development time and a quick turnaround on prototype chips than the need to reduce silicon area to a minimum, unless he is confronted with problems in partitioning between chips or is facing severe yield problems. There is thus a tendency to use gate arrays to prove chip structures and to use semicustom techniques as much as possible, reserving the involvement of the most highly skilled designers for the more difficult parts of the chip, and for cells or supercells to be utilised by the systems house designers.

The designer has needed turn round times of a few weeks for printed circuit boards: he now needs the same turn round time for chip development.

Cells that have been developed for one purpose may be combined with others to form super cells in a cumulative building process. It may be necessary to refer these cells to the specialist designer for optimisation or even redesign before using these as critical elements in larger designs.

Most MSDS designs have been produced in a three way collaboration between MSDS system designers, GEC Hirst Research Centre chip design specialists and Marconi Electronic Devices Ltd (MEDL) whose involvement ranges from detailed involvement in the actual design (and particularly designing for testability), through to a minimal involvement where they act as a silicon foundry, receiving pattern generator tapes and test procedures.

Testability is a most important aspect of chip design. Test structures must be designed in right from the start and influence the architecture of the chip. In this way we can design an hierarchy of Built-in-Test from chip, through subsystem, to system level.

The role of architecture

Cells and supercells are one way of handling increasing complexity in the design of chips. An alternative or complementary role is the use of highly regular architectures, particularly those that can exploit parallelism, so that the majority of the chip structure can be kept working all the time. In this way very good overall performance can be achieved using elements with relatively slow speed individual operation.

The use of regular structures can cut dramatically the design time and safely allow extrapolation to larger structures from small prototype structures.

One recent example is a cascadable correlator designed at HRC in 4 micron CMOS-SOS. It has 40000 transistors, 20 MHz throughput, and a figure of merit of $8 \times 10^{"}$ gate Hz/cm^2. An interesting point to note is that it took just one designer only 6 months to reach the point of pattern generation.

Another example of a regular architecture is the GRID (a device for processing two-dimensional arrays, particularly images). Here the use of a 2-D array of processing elements matched to an image scene can provide very rapid processing of the operations used in pattern recognition tasks.

An important aspect of the use of regular architectures is that they are amenable to testing.

The use of new architectures is going to require the closest possible collaboration between system designers and chip design specialists. The system designer may be able to conceive of an entirely different functional approach to capitalise on new architectures proposed by the chip specialist that can offer particularly high performance for certain classes of algorithms. On the other hand an awareness of system requirements can lead the architecture specialist to propose new architectures.

The general message is that we have progressed from the stage where the systems designer and component supplier worked in comparative isolation and the designer worked from a catalogue of standard building blocks, to a position where the systems designer, chip designer and silicon technologist are required to work in an integrated team. We have derived considerable benefit from the use of specialist courses provided by Universities such as Southampton and Brunel that help to promote an integrated approach.

Military and commercial uses of silicon Devices

The military systems designer will tend to ask for high speed, high density, low power, radiation tolerance, and operation over a wide range of temperatures. The commercial designer may ask for any subset of these (with the likely exception of radiation tolerance) but is not likely to ask for the total combination of stringent conditions.

The commercial designer may be driven towards full custom design in order to maximise yield and minimise cost for long production runs that are large compared with military requirements. However increasing complexity and the need to bring products quickly to market may eventually drive the commercial designer towards the new architectures and super cells (or microcells) already embraced by the military sector.

Packaging

Finally a plea for us to remember the importance of packaging. Much of the achievements in clever design and improvements in chip technology can be set to nought if packaging is neglected. A quick look at the ratio of area of silicon to the area of most packages illustrates some of the problems.

GOVERNMENT APPLICATIONS OF A SMALL VLSI FACILITY

E J Spall

IBM Corporation, USA

ABSTRACT

Typical constraints of government VLSI applications are described, notably very low manufacturing volumes, high reliability, high density and high performance. These constraints dictate that the semiconductor design costs must be absolutely minimised, the technology must be a stable reliable one pressed to tight lithography for high density. It is also desirable that variation part number to part number and power be minimised for higher reliability.

The trade-offs involved between gate arrays and standard cell or master image approaches, applicable for standard generic types of government requirements will be described with emphasis of the conflicting demands for high performance, low development cost and fast development times.

A 2 micron NMOS design approach, package and process line will be described for the production of such devices with typical delays of 2 nanoseconds per gate, 10,000 gates per chip and 144 to 240 pins per package. Areas to be stressed are computer-aided design, custom versus gate array and master image design approaches, make-in-house versus silicon-foundry approaches and the need for high pin-count packages.

OPEN SESSION

Titles of Presentations

VLSI REQUIREMENTS FOR RADAR SYSTEMS
P Bradsell
Plessey Radar, UK

BIT-LEVEL SYSTOLIC CONVOLVER
J G McWhirter
Royal Signals and Radar Establishment,
Malvern, UK
A P H McCabe
GEC Hirst Research Centre, UK

A SILICON VENDOR'S VIEW OF DESIGN TECHNIQUES
AND TOOLS
J D Leggett
General Instrument Microelectronics, USA

UK5000 - THE DESIGN PHILOSOPHY
J R Grierson
British Telecom Research Laboratories, UK

THE VENDOR-CUSTOMER INTERFACE
H G Adshead
ICL, UK

VLSI MANPOWER - ARE THE UNIVERSITIES DOING
A GOOD JOB?
G D Cain
Polytechnic of Central London, UK

STATUS OF VHPIC AND ALVEY PROGRAMME
W Fawcett
Alvey Directorate, UK